THE HIDDEN UNIVERSE

The University of Chicago Press, Chicago 60637

Published 2022
Printed in the United States of America

31 30 29 28 27 26 25 24 23 22   1 2 3 4 5

ISBN-13: 978-0-226-82187-0 (cloth)
ISBN-13: 978-0-226-82188-7 (e-book)
DOI: https://doi.org/10.7208/chicago/9780226821887.001.0001

First published in the United Kingdom by Witness Books, an imprint of Ebury Publishing/Penguin Random House UK, 2022.

Library of Congress Control Number: 2022932741

♾ This paper meets the requirements of ANSI/NISO Z39.48-1992 (Permanence of Paper).

THE HIDDEN UNIVERSE

ADVENTURES IN BIODIVERSITY

ALEXANDRE ANTONELLI

The University of Chicago Press

CONTENTS

Preface vii

SETTING THE SCENE: TWO UNIVERSES

How many species? 8

Unveiling the 'dark matter' of life 17

Life in peril 22

PART ONE – BIODIVERSITY: MORE THAN MEETS THE EYE

Chapter 1: Species 31

Chapter 2: Genes 44

Chapter 3: Evolution 54

Chapter 4: Functions 61

Chapter 5: Ecosystems 69

PART TWO – THE VALUES OF BIODIVERSITY

Chapter 6: For Us.....81
Chapter 7: For Nature.....91
Chapter 8: For Itself.....99

PART THREE – THE THREATS TO BIODIVERSITY

Chapter 9: Habitat Loss.....109
Chapter 10: Exploitation.....117
Chapter 11: Climate Change.....127
Chapter 12: Other Hazards and Dangers.....137

PART FOUR – SAVING BIODIVERSITY

Chapter 13: Large-Scale Solutions.....153
Chapter 14: What Can We Do?.....173

Epilogue: Looking Ahead..... 209

Acknowledgements.....215
Glossary.....217
Further reading..... 223
Picture credits.....265
Index.....267

PREFACE

Being Director of Science at one of the world's foremost plant and fungal research organisations – the Royal Botanic Gardens, Kew, in the United Kingdom – is an honour and a responsibility that I could never have dreamt of as a child. I grew up in south-eastern Brazil near the Atlantic rainforest, and it was in those exceedingly diverse, beautiful forests teeming with life that my passion for nature was born. I collected insects, seeds, shells and more, all of which I meticulously labelled, mounted and classified in old shoeboxes with a layer of Styrofoam that I had glued to the bottom.

These activities made me happy, but also frustrated, since I could seldom find the names of my findings in the translated biology books at the city library. Why didn't we know all the species that lived on Earth? That early interest and desire to seek out the concealed world of plants and animals has followed me throughout my studies and career – first in Brazil as a teenager and young student, and later in Sweden, where I moved to follow my Swedish wife, Anna. We met each other at a diving school in the Caribbean island of Utila, off the coast of Honduras in Central America, and we both became passionate about the incredible beauty of coral reefs, which we explored together while working as divemasters for some months before moving on.

After three years of travels with a tiny backpack comprising a sleeping bag, a couple of heavy dictionaries and few clothes, I decided I would do everything I could to become a biologist. So I went back to the school desk to complete my undergraduate studies in biology, followed by a PhD on the evolution of biodiversity in the American tropics at the University of Gothenburg in Sweden. Then Anna and I moved to Switzerland with our three young children, where I pursued post-doctoral studies in the evolution and diversity of plants from the southern hemisphere. We returned to Sweden in 2010, where I took up the position of Scientific Curator at the Gothenburg Botanical Garden, which holds the most diverse collection of living plants in the Nordic countries. At the same time, I began establishing my own research group, the Antonelli Lab – a diverse and collaborative group of students and researchers working across multiple disciplines of biodiversity science. I was lucky to be promoted to Full Professor of Biodiversity and Systematics at a young age, and in 2017 I founded and became the first director of the Gothenburg Global Biodiversity Centre. When I was travelling home from a sabbatical semester at Harvard University in the USA, I was contacted about applying to the position I currently hold at Kew. It was impossible not to feel thrilled at this huge opportunity, and I started my role as Director of Science in February 2019. Soon after, I was given a Visiting Professor affiliation at the University of Oxford, further expanding my scientific network in the plant sciences.

In my work, I have tried to answer the big questions about the origins and evolution of whole ecosystems, such as tropical rainforests, and how biodiversity has changed, and continues to change, over time and space – a line of research best assigned to the field of **biogeography***. I was trained as a botanist and have mostly collected plants for my research, but I've also worked on a diverse set of organisms – snakes, lizards, amphibians, birds, mammals, insects, fungi and bacteria – in order to explore and understand general patterns underlying biodiversity. I've also studied the **fossil record** (the sequence of extinct organisms preserved in rock sediments through geological time), working with colleagues to develop methods and analyses to tease out how climate change and other events in the deep history of life have affected species, and how much we can learn from the past to better predict what will happen in the future. I've had the fortune of working on and publishing scientific articles with hundreds of talented researchers, collaborating with some of them for many years.

Over time, I've also become increasingly aware of the rapid pace with which the nature I've grown to love so much, through both my early years in Brazil and my later years as a scientist, is disappearing before my eyes. The

* Words defined in the Glossary are shown in bold on their first appearance in the text – turn to page 217 for the full definitions.

scientific evidence accumulated over the past few decades is overwhelming and undisputable: we are now living in a biodiversity and climate emergency. Together, we must therefore make every effort to halt the impending disaster and loss that we are already witnessing today.

The realisation that we are living in an environmental crisis may seem dire. However, through the insights I have gained in my research I know that there is still time to fix things. As long as there are natural habitats and species left, there is hope. With knowledge about the natural world, and a willingness to care for it, come the motivation to shape our future in a more sustainable way.

This is why I have written this book. My goal is to take you on a coherent journey that begins with the fundamental basics and ends with the practical actions we all can take. In Part 1, I share my insights gained over many years of research about what biodiversity actually means, exemplifying the various components of this multi-faceted concept, and highlighting key gaps in our knowledge – which in some ways mirrors the fascinating exploration of the whole universe. In Part 2, I approach the question of why biodiversity matters, considering its many uses and values from different perspectives – practical and moral. In Part 3, I outline the major threats affecting biodiversity today, their root causes, and how these often interact with one another. Finally, in Part 4 I focus on the remaining opportunities to protect the world's declining biodiversity, from the role of our political leaders and companies, down to our personal contributions. I tend to provide examples from my own research and the work by colleagues at Kew, given my familiarity and first-hand experience with those. But I stress that these efforts are highly collaborative with other organisations and researchers around the world, for the future of our planet depends on all of us working together – and so partnerships like these are essential to our success.

Despite many years working as a researcher, professor, curator and science leader – which gave me ample opportunity to deep-dive into biodiversity – I'm just as curious and astonished at the natural world today as I was in my early years, and I have never stopped asking the most fundamental questions. Here I seek to answer these questions for you, which together explain the building blocks of all natural life on this planet. I hope that it will in turn inspire you to share my passion for our precious wildlife – our hidden universe.

SETTING THE SCENE: TWO UNIVERSES

Over a century ago, when a 30-year-old Edwin Hubble was offered a job at the Mount Wilson Observatory outside Los Angeles, he was given an opportunity that must have made all his fellow astronomers exceedingly jealous: to handle the most powerful telescope then built. One night, when he directed it towards a hazy patch of sky called the Andromeda Nebula, he made an astonishing discovery – what most people had assumed to be gas and dust was, in fact, an entirely distinct galaxy which Hubble estimated to be almost a million light years away from us. Until then, astronomers thought that everything we could see belonged to our own Milky Way, which was synonymous with the known universe.

Hubble's discovery built on the knowledge and dedicated work of many others before him, who had been equally keen to explain and measure what they saw in the skies. Some of them remain unsung heroes of science, like Henrietta Leavitt, whose mere gender excluded her from all due recognition among the exclusively male astronomers at the Harvard College Observatory where she started work in 1893. It was Leavitt who discovered a way to confidently measure the distance to stellar objects far away – an insight which allowed Hubble's estimation of the observable universe. In the years that followed, Hubble went on to map dozens of other galaxies outside our own.

In 1990, space exploration was boosted again by a space telescope which, perhaps not surprisingly, was named in Hubble's honour. It showed us in clear, colourful images that the universe is immensely larger and more astonishing than anyone could ever have imagined. Today, astronomists think there are a staggering 200 billion galaxies and one billion trillion stars in the observable universe. To put that in perspective, if every star was the size of a pea, you could cover the entire Earth with a two-kilometre-thick layer of peas. We have no idea how many planets there are, but since the existence of the first exoplanet (a planet that orbits a star outside our solar system) was confirmed in 1995, we have found over 4,300 and counting.

On planet Earth, **biodiversity** – the variety of life – is our 'hidden universe'. Its components are far more abundant, numerous, complex and interwoven than most people may realise. In Africa, our ancestors began to explore biodiversity very early on in our evolutionary history and were guided by the most basic needs of food, shelter and comfort. For hundreds of thousands of years, they tasted most things they came across. They used their senses to explore the plants growing around them and observed what other animals ate. They discovered that some plants had edible roots, but the leaves would make them sick, some plants produced sweet juicy fruits while others were bitter, and some plants were best avoided at all costs. They gradually increased their diet to include many different parts of plants, fungi, mammals, birds, fish, insects and spiders. They learned which trees provided the best wood for making fires, which animals had the best fur for keeping them warm, and which fruits were most delicious.

As our human ancestors left eastern Africa and reached other parts of the world, they made new discoveries. On every continent, the exploration of nature began immediately after their arrival. There was often abundant prey, but they had to learn how to hunt it, or satisfy themselves by scavenging on carcasses. They gathered many different sorts of nuts and berries. In some places, animals were easier to catch but fewer in number, like in the Indonesian island of Flores. The island is so tiny and food resources so limited that, when our human relatives arrived there about a million years ago, natural selection consistently benefited small individuals, leading to the evolution of a separate **species**. An adult *Homo floresiensis* was only about a metre tall (3.3 feet), which is shorter than most humans with dwarfism today and about the same size as the island's own species of dwarf elephant. In the Levant, another relative of ours, *Homo erectus* – which inhabited many parts of Africa and Eurasia for some 1.5 million years – relied on elephants, hippos, rhinos and other large animals as a source of a fat-rich diet.

As the brain size and cognitive abilities of our ancestors increased, they became increasingly better at developing tools for hunting and processing plants and animals. The development of thousands of local languages allowed them to communicate what they had discovered about the species around them and their uses. In China, archaeological evidence for herbal medicines began some 8,000 years ago. By 4,000 years ago, the Sumerians had left written accounts demonstrating their use of plants like cumin, mint and liquorice.

While most information associated with species was passed around by word of mouth, the Greeks attempted to synthesise all knowledge available at the time. In the fourth century BCE, Aristotle wrote down everything that was known to him about animals. Shortly after that, his apprentice Theophrastus did the same, but focusing on plants. Knowledge about species continued to grow, and discoveries piled up. Farmers experimented with ever more species, which increased the nutritional value of our diet and the range of climates and regions in which we could grow food. Traditional Chinese medicine – characterised by the prescription of many different plants and animals to treat all sorts of ailments – became increasingly popular across much of eastern Asia.

However, by the time western societies entered the Scientific Revolution of the seventeenth and eighteenth centuries, things had got messy. No comprehensive summaries of plant and animal knowledge had been written in Europe for almost 2,000 years, and people from different countries struggled to communicate their knowledge about species. Even when they could use the same language – usually Latin, among scientists – they didn't have a standardised way to name species. The dog rose – a common wild rose in Europe, whose fruits can be turned into a nutritious soup if you dry them and remove the hairy seeds – would be called *Rosa sylvestris inodora seu canina* by one person and *Rosa sylvestris alba cum rubore, folio glabro* by another. Not only was it a burden to learn all the different names a species could have, but also those names were often long and awkward. Misunderstandings were common and could be disastrous. In the carrot family, for instance, some of our best foods and spices – including carrots, celery, anise, coriander and parsley – could easily be confused with some of the most poisonous wild plants known, due to their similar flowers and leaves.

The person who brought order to chaos was the Swedish naturalist Carl Linnaeus. As a child, growing up in the countryside among farm animals and a rich wild flora, Linnaeus was already collecting and asking his father for help in naming everything he found – from flowers to insects and fish. He travelled the country on horseback, gathering specimens and detailing notes of every species he saw. In 1735, aged just 28, Linnaeus published the first edition of his *Systema Naturæ* ('System of Nature'), where he proposed a strictly hierarchical classification of all living things. Like the traditional Russian Matryoshka dolls, each category was contained within slightly bigger ones: species were grouped into genera, which were grouped into families, orders, classes and, finally, kingdoms. Applying this classification system, he was the first scientist to document, for instance, that whales and dolphins are more closely related to land mammals such as pigs, than to fish like tuna, despite their very different appearance and behaviour. Perhaps most importantly of all, Linnaeus proposed that every species ought to be given a single binary name: *Felis catus* for the domestic cat, *Falco peregrinus* for the peregrine falcon, and *Rosa canina* for the rose mentioned above. His naming and classification system was so simple and useful that it came as a relief to many, and, with many refinements, it has survived to this day*.

Linnaeus wasn't the only person who tried to understand and classify the plants, animals and fungi around him, nor was he the first. A particular field of science, called ethnobiology, explores how indigenous peoples have been describing, using and understanding species under other systems throughout history. Although a single scientific system for naming species has important advantages for communication and conservation around the world, this is by no means a value judgement of alternative views and practices. The reason scientists like myself have been trained in and continue to use Linnaeus's system is intimately linked to the historical legacy of colonial powers and traditional practices in higher university education

* Although Linnaeus's classification structure has remained largely intact, there have been proposals to dismiss it altogether due to the criticism that it is too rigid in its form, too cumbersome to update as new knowledge emerges, and doesn't always reflect evolutionary relationships – with critics advocating for it to be replaced by a purely genetic 'tree-based' view called 'the Phylocode'. There are pros and cons of each approach, related to stability, communication and consistency.

– and this is an important fact to acknowledge and challenge as we move forward in science.

Linnaeus's work provided inspiration for many other scholars who followed in his footsteps, beginning a new age of scientific discovery of the natural world. Several of his own students embarked on long expeditions to document the plant and animal life of places as far afield as South Africa, Chile, Australia, Japan, North America and the Arabian Peninsula. Those trips were not without perils, and several of the voyagers died young.

In London, Joseph Banks advised King George III to send British botanists around the world to find and bring back valuable plants, such as those that provided rubber and quinine. Driven by scientific curiosity but supported by the financial wealth and underlying social injustices of European empires, some of the most significant western naturalists found an opportunity to explore the natural world on their own. The German geographer and naturalist Alexander von Humboldt explored the savannahs of Venezuela and the mountains of the Andes, unveiling the tight links between geology, climate and species that still constitute the core of biological sciences and climate change research. The British biologist Alfred Russel Wallace documented the animals of the Malay Archipelago and the Amazon, revealing the striking differences in **life forms** across continents and **ecosystems** which help us understand how species cope with the environment. His contemporary Charles Darwin studied animals and plants on his round-the-world trip on board HMS *Beagle*, with a particular interest in a group of birds on the Galápagos Islands: the finches. Their beaks, he discovered, differed on each island as an adaptation to the local food source. It was from the observations made on those trips that Darwin and Wallace, independently and simultaneously, developed the Theory of Evolution – a theory that would bring to biology what Einstein's Theory of Relativity brought to physics and our understanding of the universe.

HOW MANY SPECIES?

While discoveries accumulated on the various uses of species, so did the number of known species. In Theophrastus's comprehensive book collection from *c.*300 BCE, *Peri phyton historia* (translated to *Historia plantarum* in Latin), he listed and described a total of 500 plants known to the Ancient Greeks at the time. By the end of Linnaeus's productive life, he had managed to assign formal names to some 4,400 animal species and 7,700 plant species. He never travelled further south than the Netherlands, but, based on the regular flow of specimens that his colleagues and students sent him from far-off places, he acknowledged that he had not quite covered all of Earth's species in his classification. Just before his death, Linnaeus believed that there were unlikely to be more than 18,000 species on Earth. Only time would prove that estimate wrong, by a very considerable margin.

With European voyagers bringing back specimens as souvenirs from their trips and explorers spending years in exotic countries collecting samples of everything they found, biological collections in Europe started to grow. What began as 'curiosity cabinets' for wealthy people to proudly exhibit massive shells, double coconuts and conjoined twin mammals, soon expanded to include serious and well-documented collections. Some 170 years ago in my own organisation, the Royal Botanic Gardens, Kew, in south-west London, William Hooker – the Gardens' first director and owner of the finest herbarium in private hands – filled five rooms in his private residence with **herbarium specimens** from all over the world, and three more with books. Herbarium specimens (Fig. 1) are pressed plant specimens that can include stems, and flowers or fruits attached to a paper sheet along with a detailed label; these are later identified by specialists in taxonomy and deposited in a public collection (a herbarium).

After Hooker's death, his collection was purchased by the nation, and in 1877 it was installed in the first purpose-built herbarium at Kew. As the British Empire expanded and British botanical explorations continued, a new wing had to be built almost every 30 years. This resulted in the assortment of buildings that today house what is believed to be one of the world's largest collection of pressed plants – with over 7 million specimens. Around the world, plant collections began to similarly accumulate in other national herbaria, with the establishment of what is today the largest Australian herbarium in Victoria in 1853, the largest South America herbarium in Rio de Janeiro in 1890 and the largest Asian herbarium in Beijing in 1928. Currently some 3,000 active herbaria collectively house nearly 400 million specimens. Together with other institutions hosting zoological specimens and other organisms – such as London's Natural History Museum or New York City's American Museum of Natural History – these biological collections provide our primary and most important source of information about life on Earth.

How many species are there on our planet? In this digital era, given the wealth of specimens hosted by those biological collections, you would imagine that answering this question would just be a matter of adding them up. Summing up database entries is how supermarkets know how many products they sell, and how governments know the number of people born each year. But, for flora and fauna, there are two problems with this strategy.

The first is that we haven't managed to correctly name every specimen in every collection. Many specimens were

Figure 1. A herbarium specimen deposited at the Royal Botanic Gardens, Kew. In addition to the pressed and dried specimen, the sheet contains: information about the collection location and date; the name of the collector; a description of the habitat where the plant was found; characteristics of the living plant; its scientific name (and name changes); and any other information that enriches the specimen. Additional seeds, flowers or fruits that risk being lost or are particularly fragile are sometimes inserted in small envelopes and attached to the same sheet. Digitisation efforts are now underway around the world to make herbarium specimens and their associated information free and easily accessible.

HERB. HORT. KEW.
ROYAL BOTANIC GARDENS KEW
K000430796
ROYAL BOTANIC GARDENS KEW
copyright reserved
cm 0 1 2 3 4 5 6 7 8 9 10
ZAMBIA
LAMIACEAE
North Western
Mwinilunga District; Kalene Hill, ca. 5 km N of bridge over Zambezi River on Kalene Hill-Jimbe Bridge Road, on road to Salujinga. Collection at rocky outcrop at side of road.
11°05'20"S 24°08'05"E 1320 m
Occasional; on sand and on logs; annual; stems red-purple; calyx green with red-purple spots, corolla light purple, deeper red-purple spots on upper lip inside; anthers blue.
2 March 1995
D.K. Harder, N.B. Zimba, B. Luwiika & M.M. Nawa 2854
MISSOURI BOTANICAL GARDEN HERBARIUM (MO)
Holo TYPE
of Plectranthus pulcherrimus
A.J. Paton

at first erroneously identified, and it can take decades, or even centuries, before these specimens are accurately and scientifically described and named. Many species have been given more than one scientific name – in extreme cases dozens of times (biologists love to find new things and get understandably confused when a species shows a lot of natural variation). In fact, it's often easier to describe a species as new to science than to find out if a tentative new species has already been named, because the latter requires a careful examination of all similarly looking species. Some specimens are suspected to be genuinely new to science but may not have all the features needed to ascertain this (for instance, plant specimens lacking flowers or fruits, which are often critical for telling species apart). Since species don't care about country borders, understanding how many species there are in a particular group of organisms – such as tsetse flies, chanterelles or bellflowers – requires scientists to compare lots of specimens from different regions, and often visit those places themselves to study and better understand natural variation in **form** and behaviour. That may sound like a dream job to some, but in reality it is quite tough: it requires a lot of time and funding, and gets really tricky in areas of political turmoil or disease outbreaks.

The second, and bigger problem, is that we simply don't know what's out there. Just as astronomers are still discovering new galaxies further and further away, so we find new species anywhere we look carefully enough. I've been lucky enough to travel around the world to study species in their natural habitats, and, even though my main goal wasn't to find new species, it inevitably happened. Like when I bumped into a large branch on a forest track on our first day in an expedition in Peru, which had fallen from a 10-metre-tall tree. I was going to throw it aside, when I realised it had flowers, and even fruits. I pulled out my hand lens and looked more closely at the branch, examining the arrangement of leaves and details of the flower. I quickly recognised what family it belonged to, a distant relative of coffee, but I had no idea about the species. I tucked a piece of it into a plastic bag, and when I showed it to my colleague Claes Persson later that evening (who luckily happened to be a specialist in that group of plants), he immediately knew that this species had never been named scientifically. It turned out to be a new species of *Cordiera*, a genus of about 25 species confined to the American tropics. We named it *Cordiera montana*, as a reference to the Andean mountains where the species occurs (it is now known to be from both Peru and Ecuador).

Another scientific discovery was a big gecko some 15 centimetres (6 inches) long that my students and I found in the rocky outcrops of northern Mozambique. It was after many hours hiking under a scorching sun, carrying all the food and water we needed for a few days. After the sun had set and temperatures dropped a little, we went for a walk with headtorches on. It was pitch-dark but suddenly we saw two glowing eyes stare back at us from under a huge rock. One of my students, Harith Farooq, jumped fearlessly towards it and finally managed to catch it at the cost of numerous scratches. He hadn't seen anything like it before, despite being from the region and knowing the local lizards like the back of his hand. It was a remarkable gecko in many respects: possibly the largest one in Mozambique, with beautiful colour patterns over its body, big yellow eyes, a ring around its nostrils and very fragile skin, which would peel off at the slightest touch, as a strategy to escape predators. And like us on that expedition, they slept through the day and were active in the marginally cooler night. However, it is still not clear whether it is a new undescribed species, or a previously unknown population of a known species – *Elasmodactylus tetensis* – found hundreds of kilometres away from where we were; either way, this was an exciting discovery.

In the tropics, finding new species is not that difficult – if you know your species, know what you're looking for and go to places where few biologists have been. A case in point is American botanist Charlotte Taylor at the Missouri Botanical Garden, an internationally prominent organisation in plant sciences and conservation. Also working with South American plants, as I did for my PhD, Taylor is one of the most prolific botanists still active today. She has described about 500 plant species and set names to some 400 others, which had previously been known but needed to be re-classified (for instance, by finding evidence that they belonged to a different genus than originally thought). Carrying out fieldwork with someone so knowledgeable, as I was fortunate enough to do while in South America, is an amazing experience.

But even in heavily studied countries like Sweden, the birthplace of Linnaeus, you could be lucky, especially if you have a passion for obscure, underappreciated creatures. In 2007, about a dozen scientists from several countries were invited to spend two weeks on the beautiful island of Tjärnö, where the University of Gothenburg has a research station. The goal was simple: to look for new species of tiny creatures around the station. All their expenses were paid for, and they were given access to boats equipped with dredges for taking sand and mud samples, and everything else they needed. They found an astonishing 27 unknown species. These included 13 new species of copepods – relatives of shrimps and crabs that are abundant in every sea and lake.

Given so many unknowns and constant discoveries, it is perhaps not surprising that the total number of species on Earth is, at most, what we can call an informed guestimate. There are currently *c.*3.5 million species scientifically described. Of these, scientists believe that about half are synonyms – those described more than once and so having two or more names, in which case only the first description and name is considered legitimate. This leaves a total of 1.8 million 'valid' species. From there on, it's anyone's guess. In the early 1970s, some American scientists set up a large blanket under trees of a single species in the Panamanian rainforest and released some pretty nasty gas towards the canopy to see how many different bugs would fall dead. From that single tree species, they found almost a thousand species of beetles.* Although the ethics of such sampling are questionable and there are less destructive ways of doing such work today, by then the work was well received and made big news.

Today, many biologists I know seem content with the estimate that *c.*8.7 million species live on land and in the sea. But this may be more a sign of their lack of interest in speculating about something no one can really prove right now. One thing is certain: this number will change, and will most probably increase. Technological developments over the last few decades have allowed us to detect ever smaller species, and those that are very rare or specialised. We're now assessing the biodiversity of places we couldn't previously access, from thermal vents in the deep ocean to the dense forests of Papua New Guinea. As scientists, we are also increasingly encountering unknown species in our own biological collections, such as fungi living exclusively within the seeds of certain

* They multiplied that number by the number of known tree species to predict a diversity of 30 million insects in the tropics alone. However, this extrapolation was based on certain questionable premises, such as the assumption that a large proportion of insects were only able to eat the leaves of one particular tree species and that the number of insect species was therefore directly correlated with the number of tree species. In other words, scientists initially thought that most insects were extreme specialists: for example, if a grasshopper species did not find leaves of the 'right' bush to feed on, it would die rather than eat something else. Since then, we've come to understand that insects are far more generalist than that, bringing the total estimate down by at least ten times. Many other studies have attempted to extrapolate species diversity, inevitably basing their estimates on incomplete information.

plants, or the lichens and mosses on leaves and branches of herbarium specimens.

Further, the 8.7 million estimate excludes a significant and substantial portion of all diversity: bacteria and **Archaea**. For both of these groups, species' boundaries and definitions are less clear. Once you include these two groups, serious calculations suggest that a trillion (!) species might in fact share this planet with us. As a comparison, the Milky Way is estimated to contain some 100-400 billion stars. This shows the level of discovery and understanding that lies ahead of us.

If figuring out how many species there are today wasn't enough of a challenge, to really understand biodiversity we must not forget to look at species that have already gone extinct. We do this by looking back into the fossil record. The millions of fossilised specimens found around the world, combined with statistical models that predict how many have not yet been sampled, suggest that about 99.99 per cent of species that ever lived are already gone. It is therefore obvious that we have barely scratched the surface in our understanding of life on Earth.

UNVEILING THE 'DARK MATTER' OF LIFE

Astronomers have one piece of technology to thank for the leaps they have made in exploring the universe: the telescope. So do biologists, but our truly transformative technology is not the microscope, it is DNA sequencing – the laboratory technique used to determine the sequence of bases (A, C, G and T) in a DNA molecule. DNA, the molecule that contains the biological instructions to build and maintain an organism, was discovered during the lifetime of Darwin – but it wasn't until 1953 that its double-helix structure was clarified, and much later, in the 1990s, that it started being used in biodiversity research. From the beginning, this was a major undertaking: it took 13 years and an estimated US $2.7 billion to produce the first human **genome**. Today, you can get yours sequenced in a couple of days for less than US $300 (£200). Or you can pay even less if you're satisfied with sequencing a smaller portion of your genome, enough to reveal a great deal about your ancestry and – if you wish to know – your predisposition to certain diseases. Among many discoveries, DNA sequencing technology has helped clarify our position among all living organisms alive today, just as telescopes helped find our position in the universe (Fig. 2).

Routine DNA sequencing has opened up previously unimaginable possibilities for identifying species based on genetic differences. A few years ago, a Brazilian PhD

Figure 2. Finding our place. *Left:* The Hubble Sphere, a schematic view of the universe from Earth's perspective at the centre, to the current limits of observation, 46.5 billion light years away. *Right:* The Tree of Life, depicting how all life forms on Earth share a common ancestor some 3.5 billion years ago, built using differences among the DNA of living species.

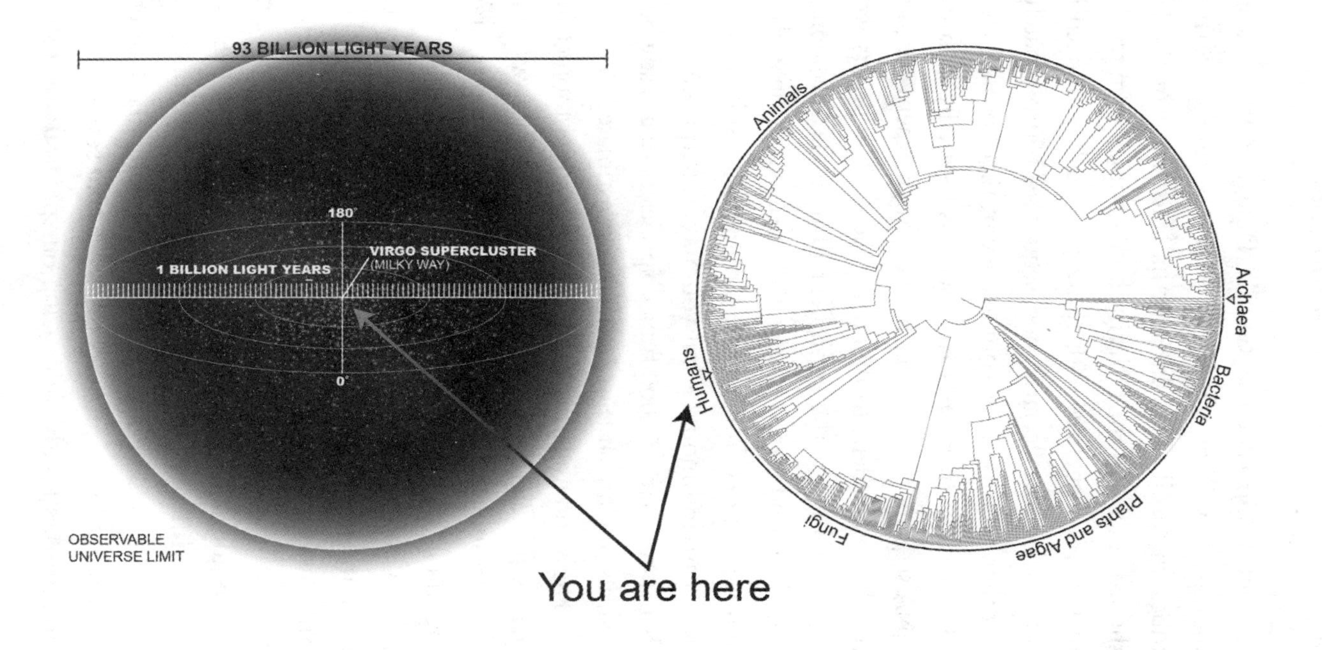
93 BILLION LIGHT YEARS
180°
VIRGO SUPERCLUSTER
(MILKY WAY)
1 BILLION LIGHT YEARS
0°
OBSERVABLE
UNIVERSE LIMIT
You are here
Animals
Archaea
Bacteria
Plants and Algae
Fungi
Humans

student I supervised – Camila Duarte Ritter – spent months travelling across the Amazon, filling up small tubes with soil from different habitats. Back in the lab, she sequenced all the DNA in those samples, then tried to match her sequences with those previously produced by other scientists. Because most sequences had no matches in other databases, she couldn't assign them proper scientific names. In such cases, scientists usually consider DNA sequences that differ by more than 3 per cent to be roughly equivalent to different species. The results were jaw-dropping. In the equivalent of a single teaspoon of soil, she found up to 1,800 'genetic species'. Of these, about 400 were fungi. You will be familiar with shiitake, truffles, yeast and mould – but you may be surprised to hear that more than 150,000 fungal species have already been described, while at least 3 million are estimated to exist.

High and underexplored biodiversity is not something that is peculiar to tropical rainforests. In fact, we don't need to look far to see equally striking examples, starting in our own bodies. On our skin and hair and in our cavities and gut, healthy individuals are home to over ten thousand species of microbes, a large proportion of them still undescribed scientifically – bacteria, archaea, fungi and viruses. Their cells outnumber ours several times over. In our gut alone, our co-inhabiting bacteria host over two million different genes, which is about 100 times more than contained in our own DNA. This human ecosystem, our **microbiome**, provides numerous but scarcely known **functions**, strongly influencing our physical and mental health, our immune, endocrine and nervous systems, and leading to – or preventing – a wide array of illnesses, from inflammatory bowel disease to cancer and depressive disorders. We inherit a large proportion of our mother's microbiome during or shortly after birth, and within our first year of life some 1,000 species colonise our gastro-intestinal tract, leaving a lifelong microbiome signature. The microbiome varies widely among individuals and age, often increasing in diversity as we get older. There are large differences among individuals and regions, with strong links to our diet. Medical treatments, in particular the ingestion of antibiotics, can largely disrupt the system, but the microbiome eventually returns to a state of equilibrium, even though its species composition may change.

Physicists and astronomers are still looking for what they call 'dark matter' and 'dark energy' – unseen cosmological components needed to describe the observed dynamics of our universe. Together, cosmologists believe dark matter and dark energy account for 95 per cent of our universe, but in fact, they are struggling to truly understand what dark energy and matter are. Similarly, biologists are just at the beginning of a long journey in understanding biodiversity. Linnaeus used to say that you can't understand what you can't name. Finding and naming all species is therefore a critical step, and one we're still very far from achieving. That task could take centuries, unless current methods of describing species are improved and accelerated, and public investments increase substantially (in which case this could instead take some 50 years of intensive research). But it's really just a first step in a much longer journey: naming species is our springboard to understanding their role in the environment, their benefits to other species and humankind, and what might happen if they disappear, or start to multiply uncontrollably.

LIFE IN PERIL

There's much that connects astronomy and biology: the wonder and beauty of discovery, the large unknowns, and the fact that we're all just 'star dust' – as nearly every element on Earth was created in the heart of stars. I'm equally fascinated today when I learn about the amazing interplay among species as I was when my father told stories of the galaxies and planets during our stargazing nights outside my home city Campinas in Brazil. But, there is one crucial difference. The stars I looked at as a child are essentially the same as today. The forests that surrounded our night adventures, however, are all gone.

Throughout our exploration of life on Earth, we have not been satisfied with observing, learning and carefully profiting from the species around us. Instead, we have left profound and often irreversible destruction. If our planet's entire 4.5-billion-year history could be condensed into one day, modern humans would have made their arrival six seconds before midnight. Yet, these few seconds – which translate into some 300,000 years of *Homo sapiens* – have seen the world transformed in ways so radical that it is almost impossible to grasp. In the cradle of humankind in eastern Africa, we set fire to vast swathes of savannah to help us spot and chase game, a trick we later took with us into Europe, Asia, Australia and North America. In South America, thousands of years before the first Europeans set foot on the continent, people in the Amazonian rainforest had been hunting monkeys and rodents, moving around their favourite plants – cacao, manioc, Brazil nuts – and clearing large chunks of what has long been thought of as one of the last untouched havens of nature.

The traditional use of natural resources by indigenous communities around the world has been sustainable in many ways, but the same cannot be said about the changes that came much later. It wasn't until the 1950s – 1.3 milliseconds before midnight on our world clock – that the gradual transformation of the planet turned into something completely different: a massive, ubiquitous, and disastrous transformation of nature. Even though the drivers of change had existed for a long time before, for most regions their intensity underwent a drastic acceleration. In these few decades, we've lost a quarter of all tropical rainforests, pumped 1.4 trillion tons of the planet-warming gas CO_2 into the atmosphere, and added over 5 billion people to the planet.

As a consequence, species are now probably disappearing faster than any time in human history*. On

* While I believe this often-quoted statement to be true, this is a contentious debate among scientists. A key challenge is that it's notoriously difficult to prove a species to be extinct: absence of evidence is not evidence of absence.

every island, every continent, every coral reef, a significant proportion of the world's species are becoming increasingly rare, until they will one day disappear, never to return. Several hundreds of species of mammals, birds, plants and frogs that were still alive in the 1500s have been confirmed extinct, while the true number is certainly many times more. Today, scientists estimate that about a fifth of all species may face extinction in the coming decades. If this happens, it could be classed as a new wave of 'mass extinction' – a period when species extinctions are exceedingly larger than the normal background rate. This planet has only had five generally agreed mass extinction events in its long history – the last one 66 million years ago and caused by the impact of a 12-kilometre-wide asteroid that hit Earth off the coast of Yucatán in Mexico. Today, humans are becoming the equivalent of that asteroid.

It is no exaggeration to assert that the escalating degradation of nature and associated loss of biodiversity pose an existential threat to our own future. As the species around us disappear, we lose invaluable sources of food, medicine, fibre, clothing and many other properties that we have barely begun to explore, but which could provide solutions for the next pandemic or for ending hunger. As deforestation continues in Amazonia, the whole ecosystem now risks passing a tipping point, after which large regions may irreversibly convert into savannahs – greatly reducing regional rainfall and water supplies for tens of millions of people and releasing vast amounts of greenhouse gases that drive global climate change.

Despite the huge challenges facing biodiversity today, there is still time and there are concrete ways to reverse the negative trends of global extinction and local losses. But action requires commitment, and commitment is best brought about by an emotional connection – a deep sense of meaning and purpose that we can only gain from our personal experiences. I am sure you will have your own fond memories in nature. I don't think that any person could remain emotionally unaffected while standing in front of a monkey carrying her baby on her back as she grabs and eats a banana you give to her; or at an island cliff packed with noisy and restless seabirds; or in a desert in full blossom after the first rain in years. No matter how many nature documentaries we may have seen from our sofa, no one can be really prepared for the things that happen in reality, when we are not merely observers but truly part of it. Instinctively, we are all trained 'biologists' and 'life-learners'. I believe that keeping this inbuilt condition alive, through our lives and that of our descendants, is the most critical factor in building a world where humankind and nature can co-exist and thrive.

Hubble not only discovered that there were other galaxies like our own Milky Way, but also that those 'island universes' appeared to be speeding away from us. The more recent cosmological discovery of dark energy shows the universe is expanding at an accelerating rate, probably even faster than the speed of light. If confirmed, this would mean that today's farthest observable galaxies will speed away so fast that they effectively disappear. Just as our time for discovering and mapping biodiversity on Earth is limited, so is our time for cosmological discovery of those galaxies. We can't change the dynamics of the universe, but we are already changing the dynamics of our planet – and not in a positive way. The good news is that we can – still – stop the degradation of our natural world. But if we are to care for biodiversity, we first need to fully understand what it is.

PART ONE

BIODIVERSITY: MORE THAN MEETS THE EYE

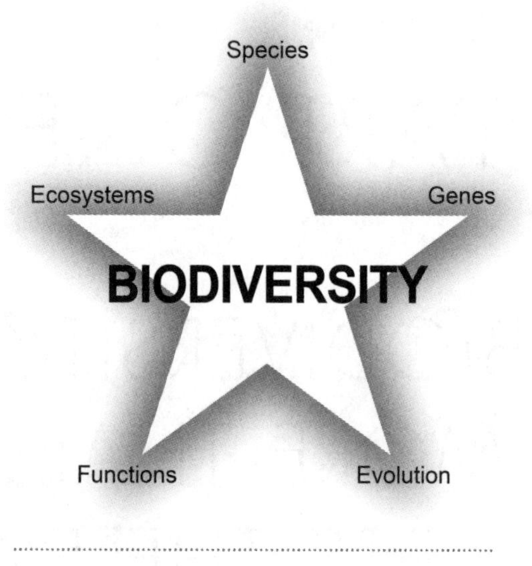

Figure 3. The Biodiversity Star. The concept of biodiversity encapsulates multiple complementary, but distinct, facets – each of them depicted here on a separate apex. Identifying, measuring and monitoring these five aspects of biodiversity is crucial to our understanding of life on Earth and our ability to take action when one of them begins to fall apart.

Just as the universe is turning out to be far more complex than we first thought, so is the biodiversity on our planet. Its breadth and depth are far greater than we realise. The term 'biodiversity', a contraction of biological diversity, was brought into wide use by American biologist E.O. Wilson. At its simplest, biodiversity is 'the variety of life on Earth'. However, biodiversity is in fact a deeply intricate concept that encompasses a plethora of characteristics and meaning. I think of it as a 'five-pointed star' (Fig. 3). Each point is linked, but distinct. They are: species diversity; ***genetic diversity****;* ***evolutionary diversity****;* ***functional diversity****; and ecosystem diversity. Alone, each apex cannot convey the full variety of life – you need them all, in the same way you need all your fingers and not just your thumb to be fully efficient. Conserving just one of these factors, such as the number of species in a given place, could sacrifice disproportionately one of the others, such as the species' evolutionary or functional diversity. In this section, I will explore how each component of the biodiversity star interlinks: what they mean, why they matter, and how together they drive a healthy and resilient living planet.*

CHAPTER 1

SPECIES

Species are the bedrock of the living world. They are equivalent to the bricks of a house, the elements of the periodic table, the keys of a piano. All species have a place in nature; they exist in communities that depend on each other and their physical environment. Yet despite their vital importance, scientists cannot agree on a single definition of what a species is.

The classical concept of a species is a group of individuals that can breed and produce fertile offspring with one another. Largely sterile offspring are therefore not considered separate species, such as those resulting from rare sexual encounters between lions and tigers at zoos (called 'ligers' for those with a lion father and tiger mother, or 'tigon' for the opposite). In contrast, if you cross a poodle with a Labrador, their puppies will be just as fertile as the parents: therefore, all dogs belong to the same species. Even wolves can successfully breed with domestic dogs and produce fertile puppies; therefore, wolves and dogs belong to the same species, *Canis lupus*. That said, compared to wolves, all dogs share many commonalities, such as shorter teeth and tameness. Those shared characteristics are signs that dogs may be collectively drifting away from their wild ancestors and may, eventually, become a separate species. Dogs are therefore considered, today, a **subspecies** – *Canis lupus familiaris*.

In practice, the standard definition of a species doesn't always work. So far, no one has tried to cross a hippopotamus with its closest living relatives (the whales) to test whether they can interbreed! We therefore need other ways of telling species apart, such as looking at their genes. If a biologist like myself suspects that we have encountered a species previously unknown to science because, for instance, one or several individuals look or behave a bit differently to others, we extract a small amount of tissue from them (such as a blood sample from an animal, or leaf fragments from a plant) and sequence portions of their DNA. Using various methods and computer programs that estimate the most probable evolutionary tree based on differences in DNA sequences, we are then able to find out whether the seemingly unusual individuals form a genetically distinct cluster – a group that has different genetic characteristics from other known species. If this is the case, it would provide evidence that they are isolated and have not bred, and have therefore not exchanged genes with other species. Bingo – we have a new species on our hands.

This is what happened not so long ago in one of the most-studied countries for its biodiversity, the United Kingdom (where I spend most of my time), and for one of the best-studied **organism groups**, mammals. In 1993, researchers using bat detectors (audio recorders capable of capturing sounds at higher frequencies than we are able to hear) were surveying the bat fauna around Bristol, when they realised that some individuals of the most widespread British bat, the common pipistrelle (*Pipistrellus pipistrellus*), were emitting calls at a slightly different frequency (55 kHz) than the researchers were used to recording (45 kHz). They captured some individual bats and soon found out that what they had documented was in fact a different species, which was formally recognised in 2003 as *Pipistrellus pygmaeus*, the soprano pipistrelle. Besides the consistent difference in echolocation calls, further studies also showed small but clear differences in skull form, behaviour, and perhaps most importantly, DNA. The fact that a mammal species had been overlooked for over 200 years – since *Pipistrellus pipistrellus* was first described scientifically in 1774 – was a very big deal, particularly as only 16 bat species were known to breed in the country. It later turned out that the new species was not only very common, but also distributed widely across Europe.

DNA is also helping to add to the list of native fungi found in the United Kingdom, which is currently increasing by 50 or more species each year, including some previously unknown to science. When most people think of fungi, they think of mushrooms, but these are just the fruiting bodies of a much larger organism growing within the substrate beneath – in soil for instance or a fallen log. The 'mushrooms' are just like apples on an apple tree, with the difference that the tree is practically invisible to us. They pop up only sporadically, sometimes not at all, and represent only a tiny fraction of the total weight of the fungus, which is primarily made up of fungal threads called hyphae, grouped together in a network known as a mycelium. Sometimes they can get really, really big. In fact, the world's largest organism is not what you may think: it is the honey mushroom in the genus *Armillaria*. The hyphae belonging to a single individual in Oregon in the United States have been found to possibly weigh as much as 35,000 tons – equivalent to 250 blue whales – and this individual is some 2,500 years old. The predominance of fungal tissue embedded in the environment is also why, nowadays, fungal surveys are often carried out by simply taking samples of soil across a forest or meadow, bringing them to the lab, and examining how many different entities the DNA sequences reveal. However, to put a name on the sequences, you need to match them against a set of references, produced from confidently identified museum specimens. This matching exercise is known as 'DNA barcoding', given its similarity to how products in a supermarket are identified at the checkout.

DNA can now solve most problems of species identification, but not all. One important exception is for species that have gone extinct. DNA invariably degrades with time, and does so faster at higher temperatures. Even under the best preservation conditions, there is an upper theoretical 'expiry date' of some 1.5 million years, after which all its pieces will have broken apart. (This is why the science behind *Jurassic Park*, where the scientists extract dinosaur DNA from mosquito bellies preserved in 80-million-year-old amber, is nonsense unfortunately – or perhaps fortunately, considering how things went in that story!) Some of the oldest DNA fragments sequenced so far are from the teeth of a mammoth buried in the Siberian permafrost (soil that is permanently frozen), which are just over one million years old. In contrast, nearly all extinct species have left no DNA behind. To tell fossil species apart, you therefore need to carefully study their form, and sometimes make assumptions of how feasible it is that similar-looking fossils separated by large distances or time could belong to the same species. This gets even trickier when you only have a tiny part of the whole organism to look at – like fossil pollen, or imprints of leaves – and nothing else to compare with. Several colleagues, including Carlos Jaramillo and Monica Carvalho in Panama, and Carina Hoorn in the Netherlands, have unveiled incredible insights by carefully studying such material.

Another challenge is when members of a presumed species look the same, have similar DNA, but seem too far apart to ever reproduce with one another naturally. Some years ago, I collaborated with a PhD student – Lovisa Gustafsson – who bravely travelled to three Arctic regions thousands of kilometres apart: north-western North America (Alaska and Yukon), the archipelago of Svalbard in the North Atlantic, and mainland Norway. In each region, she collected tens of plants from the same set of species. She brought all of them back to a greenhouse at her home institution in Oslo, Norway and painstakingly pollinated them by hand to see if they were able to produce fertile offspring. To everyone's surprise, five of the six species that survived the whole experiment could not reproduce successfully across populations. This suggested that there are multiple **'cryptic' species** hiding within these five names, all looking alike but unable to reproduce with plants from different locations. In other words, what looks like just one species is likely to be several different ones. No one knows yet how widespread this phenomenon is, but if common it could mean that we're grossly underestimating the number of species in the Arctic, and perhaps in other regions too.

The opposite of this pattern is when species *can* reproduce successfully, but don't normally do that. Among plants, orchids are notorious examples. With some 28,000 known species and many more documented by scientists each year, the orchid family is one of the two largest plant families in the world. Why orchids are so diverse has puzzled scientists for centuries, including Darwin. An important insight came from the discoveries of the Australian naturalist Edith Coleman. In 1927, she published detailed observations of how a species of tongue orchids – *Cryptostylis leptochila* – was pollinated by wasps belonging to the species *Lissopimpla excelsa*. The curious thing was that the insects were not getting any food reward for the task of moving pollen from one flower to another. In fact, what she deciphered was highly peculiar: the wasps behaved just as if they were... well, mating with the flowers. It turns out that the flowers emit scents that are identical to those produced by female wasps. What is more, the tiny hairs on the petals enhance the sexual act, and the male's eyes cannot differentiate between the colour of the flower and the colour of female wasps.

The phenomenon Coleman described for the Australian species – now known as 'pseudocopulation' – is by no means an isolated case. Today, one in three orchid species is known to deceive insects in one way or another, often by harnessing the sex appeal of females. I witnessed this first-hand while doing a pollination study for one of Europe's most spectacular orchids, the lady's slipper orchid. I had two weeks of holidays between the end of my MSc and the start of my PhD, so a friend and I thought we could use the time to get some research done. We were struck by how male bees were effectively lured into the orchid flowers time and again, without seeming to ever learn the lesson that there was no reward for them.

Mastering the trick of deception has required many orchid species to become highly specialised to their pollinators, since the chemicals, shape and colours that attract one insect species are often very different from those that attract another. Although each orchid species is distinct in form and DNA from all others, if you manually took pollen from one species and added it to the female reproductive parts of another, chances are you would get a perfectly fertile crossing. Were its parent plants indeed different species? Most botanists would say so. Even though this may seem at odds with the classical species concept, those species have mechanical barriers that, in nature, prevent them from interbreeding.

Strongly linked interactions between plants and their pollinators are not restricted to orchids. At the most extreme, a single species of bee, fly, bird, or another animal, has a body shape that perfectly fits the flower shape of one specific plant species. This allows the animal – and often only it – to reach the floral nectary to get its reward, and in doing so it pollinates the plant. High levels of pollinator specificity are most common in the tropics, which is thought to be one of the reasons there are so many species there. In cooler regions, pollinators are more often generalists – feeding on, and pollinating, various species and even families of plants. In the example of the lady's slipper orchid mentioned above, which grows in temperate regions, at least a handful of bee species are able to pollinate it, although some more effectively than others.

Another case where DNA doesn't provide all the answers, and in fact can just make things more complicated, is for species that remain distinct in form despite occasionally exchanging genes with others (which they shouldn't, by definition). This happens a lot among bacteria, which cannot be easily shoehorned into any sort of strict species delimitation. Some plants are also known to exchange genes across species, such as bromeliads (relatives of the pineapple) in Brazil's Atlantic rainforest. Those events could perhaps be due to 'mistaken' pollinations, such as when a pollen-carrying insect or bird flies into the flower of a different species than they usually do. Among mammals, horses and their closest relatives (zebras and asses) provide yet another intriguing example – with genes being exchanged among several independent species through their evolutionary history, despite striking differences in their form, and even different chromosome numbers in their cells. And we don't have to look far for further examples: even *we* have received genes from other species. Today, about two per cent of the DNA in all non-African humans derives from our extinct relative *Homo neanderthalensis*, showing the legacy of occasional sexual encounters. In other words, only those human lineages that left Africa early in our history encountered – and mated with – Neanderthals.

THE GEOGRAPHY OF LIFE

Despite the challenge that there is no single, universal criterion to identify species, scientists like myself are working hard to find out where each species exists on our planet. Imagine we could overlay an imaginary grid of equally sized squares across the Earth and record the presence and absence of every species within it, both on land and at sea. The total number of species present in each place – the **species richness** – would help us prioritise which regions we ought to protect, and where we could build the next urban expansion or establish a new crop field without causing a major impact on biodiversity. It would also tell us where species currently facing extinction live, so that we could better protect them in their natural habitat.

Alas, to do this globally is not an easy task – it would require sending experts in multiple organism groups to every corner of the Earth. Instead, much of the information we have gathered so far about species is serendipitous. If you look at a map of all the species collected or observed in Australia over the last decades, you get an almost perfect map of the country's roads. This is not because most species particularly enjoy roads, but rather because those making observations are far more likely to do so along roads than in areas that are difficult to access. Over time, this has led to a highly patchy and biased understanding of species occurrences. Even so, we know quite a bit about the geography of life, with three main insights emerging.

First, not all species are everywhere: they all have specific tolerances in relation to the environment. For instance, fish living in the deep sea cannot cope with the pressure of shallow waters, and mammals in the African savannah wouldn't survive the Siberian winter. Their current distribution is also the result of historical legacies and geographic constraints, such as penguins in Antarctica not having had the opportunity (yet) of colonising similar habitats in the northern hemisphere. The location and size of different regions are factors that co-interact in defining not only which species occur where, but also how many species are formed. The **'species-area relationship'** is one of the few 'rules' in biogeography, and is an important component of the Theory of Island Biogeography (Fig. 4) published by American ecologists Robert MacArthur and Edward O. Wilson in 1967. The theory predicts that the

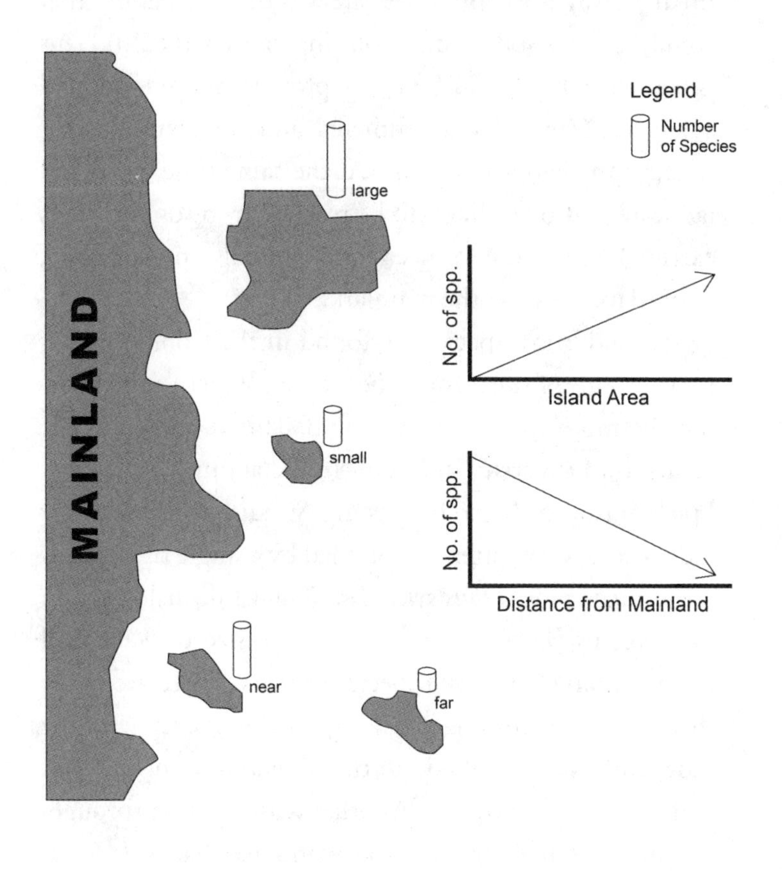

Figure 4. Predictions from the Theory of Island Biogeography. Theory and observations match each other remarkably well in determining that the number of species on an island can be predicted from the size of the island and its distance from the mainland, as depicted by the simple linear plots shown here.

number of species on an island or in an island-like environment (like a forest fragment amidst savannahs) will increase with the size of the island, and will decrease the further away it is from other areas of similar habitat that could act as a source of colonising species (Fig. 4). This is because a big island generally provides more food, and a larger variety of environments and opportunities for **speciation** than a small one. At the same time, an island far away is also less likely to be reached by birds and seeds carried by the wind, or occasional animals on driftwood, than islands just off the mainland.

Second, most species are found in the tropics. Across most groups of organisms, the closer you get to the equator, the more species you will find. This is known as the **latitudinal diversity gradient**. In the autumn every year, I pick mushrooms and berries in a Swedish forest with my kids. There, we're often surrounded by a single tree species – the Scots pine, *Pinus sylvestris*. When I do fieldwork in Ecuador, in contrast, an area just the size of a football pitch contains up to 500 species of trees. There are many theories to explain these striking differences; some of the most credible ones allude to the higher amount of water and energy in the tropics. Another reason is that throughout most of life's history, the world has had a tropical climate, giving tropical organisms more time to speciate into myriad life forms than those in cooler zones.

Third, most species are rare, either naturally or due to various human impacts. One of the Brazilian PhD students I've co-advised, Maria do Céo Pessoa, was studying a group of plant species distributed across South American rainforests. One day, she found a plant specimen collected in a reserve near Manaus, at the heart of the Amazon. It had been identified as *Chomelia estrellana*, but after careful examination she realised that was a mistake: in fact, it was a *Chomelia triflora*. The problem was, the nearest tree of that species had only been found in French Guiana, over 1,100 kilometres away. That sounded very unusual to us, but was it really? I suggested we teamed up with some colleagues and another student, Alexander Zizka from Germany, to investigate quite how uncommon it was for plant species in the American tropics to be so rare and so far apart from each other. To our surprise, we identified over 7,000 other species only known from just two locations, and, of these, one in five had populations that were more than 580 kilometres apart. Our work, and that of other researchers, has consistently shown that being rare is, paradoxically, very common. Mapping rare species is critical, to ensure their protection, but the task is like finding needles in a haystack.

Identifying and figuring out the distribution of species is crucial for understanding the natural world, but it is not enough. Species – unlike bricks, atoms or keys of a keyboard – vary drastically among themselves. The main source of this variation is what we find within them, in each one of their cells: their genes.

CHAPTER 2

GENES

Imagine Earth was about to be hit by a wayward asteroid and some of us were lucky enough to escape aboard a spacecraft. If we wanted to take some plants with us, how many seeds of each species would we need? Would one or two be enough, or are there such differences in form and function that we would need many more to capture their essence?

Genetic diversity, the variation in this DNA blueprint, plays a role in determining the pedigree of elephants and tigers in the captive breeding programmes of zoos and game parks. Within each species, there is often a great deal of variation in form, chemistry and behaviour, all linked to the organism's DNA, or the 'blueprint of life'. For the vast majority of species, however, we have little clue about how much genetic variation there is, and which genetic variants within a species may be best at withstanding the environmental changes taking place around the world today. As the world's climate becomes warmer and more unpredictable, it's crucial that we understand how each species will be able to tackle those changes, including those species on which we depend, such as crops.

A compelling case that illustrates this is coffee (Fig. 5). Today, some 15 million farmers in Ethiopia rely on growing the commodity, which is one of the country's main sources of export income. With temperatures set to increase this century by some 1.5–5.1°C (2.7–9.8°F), depending on the emission scenario, researcher Aaron Davis and colleagues at Kew and in Ethiopia have predicted that some 40–60 per cent of currently cultivated areas may no longer remain suitable for the crop by the end of the century, with some serious production shortfalls in the next 50 years. This would lead to serious socio-economic challenges, such as income and food insecurity, substantial migration and even conflict. Cooler and wetter areas at higher elevations could provide more suitable growing conditions, enabling substantial production of coffee even under climate change. However, this would require a massive and complicated migration of the Arabica coffee plantations, risking conflict, deforestation and a loss of opportunities for farmers elsewhere. Davis and his collaborators are responding to this dilemma by locating coffee plants in wild populations that have useful genetic traits, particularly those that are more resilient to heat and drought. They are also exploring the flavour, cultivation potential and climate profiles of several lesser-known species in the coffee genus, which includes some 130 species and counting. These could replace more common coffee crop varieties – such as arabica and robusta – or be interbred with them, in order to dramatically increase their **climatic tolerance**, and in turn help to preserve the livelihoods of millions of people worldwide.

Gene-editing technologies now allow for the transfer of particular DNA fragments across species boundaries. In a famous experiment in the 1990s, researchers in California were looking for a way to give tomatoes resistance to frost, a common problem for outdoor farmers. They came to think that Arctic fish might hold the solution, since they can live in extremely cold waters without their blood freezing. They managed to extract a gene responsible for producing proteins that inhibit ice crystallisation, but when they transferred it to the tomatoes, it just didn't produce the same effect. The trial was abandoned, and the tomatoes never made it to the market. Since then, gene-editing technologies have improved enormously, with today's focus now geared towards identifying useful genes among the wild relatives of crops.

I was personally against genetically modified crops early in my university studies in biology, since I thought that resources were better spent elsewhere and there could be potential negative effects for the environment. But I have since changed my position. If done responsibly, the evidence for negative effects on the environment or human health are almost non-existent, as compared to the very real and substantially documented problems that such crops aim to address, such as biodiversity loss and food insecurity. Examples of gene editing carried out today include attempts to improve a plant's ability to use

Figure 5. Flowers and fruits of Arabica coffee. This is one of the world's most traded commodities, and a major income source for many tropical countries around the world. Its cultivation is now being threatened by climate change, prompting scientists to find other species or varieties that tolerate higher temperatures and more unpredictable weather conditions.

soil nitrogen, in order to increase yield per land area and reduce the demand for converting natural habitats into cropland; and inserting genes into common crops so that they produce vital micronutrients not easily accessible by people in many low-income areas of the world.

One example of natural genetic variation that may prove very useful is in the European ash. This species plays a critical role in ecosystems, with more than 1,000 species associated with it, including over 50 mammals and some 550 species of lichens. Unfortunately, the tree is now threatened by the ash dieback disease, caused by a fungus previously only found in parts of Asia. Since the fungus was first discovered in Poland in 1992 (probably derived from imported plants), it has been spreading westwards and is found today in most European countries. Millions of trees are now heading for certain death, and there is no known cure. In the United Kingdom for instance, ash is the second most abundant tree species, and 80 per cent of all trees are expected to die in the near future, which is expected to cost the country nearly £15 billion (US $19 billion), from felling dead trees and planting new ones, to the loss of key **ecosystem services** they provide. This showcases the fragility of ecosystems and the economic losses associated with the spread of pathogens. Luckily, however, after sequencing the DNA of thousands of ash trees, Kew scientist Richard Buggs and collaborators have found that a small fraction – less than 5 per cent – are naturally resistant to the fungus. Propagation of those individuals now gives hope that future ash woodlands won't succumb once again. The ash dieback fungus is just one among hundreds of other agents of plant and animal disease where genetic diversity can come to the rescue.

Unpredicted changes have certainly become more common due to human activities, but they arise even under fully natural conditions. For four decades, the American biologist couple Rosemary and Peter Grant spent six months every year studying the finches of Daphne Major, a tiny island of the Galápagos archipelago that gave Darwin the core ideas for his evolutionary theory. By carefully tracking, measuring and observing every single finch on the island throughout that period, the Grants were able to witness first-hand how tiny, genetically linked differences in the size of beaks played a major role in the fate of individuals. When El Niño events brought an excess of rain that led to a shift in the island's vegetation from large, hard seeds to smaller, softer seeds, this benefited birds with smaller beaks, which were then more efficient at eating, and produced larger offspring. A subsequent drought year, in contrast, would turn things around and benefit those that have bigger beaks and could crack the tougher seeds.

Over time, these examples show that it is genetic diversity that makes species better capable of dealing with change. It's like going camping and bringing a Swiss knife with lots of functions, rather than one with only a single large blade; you just cannot know in advance what you may need. Similarly, in many cases, it's not possible to assess how much genetic variation a particular species has by only examining it visually. Consider, for example, our own species. Despite our amazing variation in culture, look, language, religion and traditions, everyone alive today shares 99.9 per cent of their coding DNA with each other. In other words, humans have a tiny amount of genetic variation. This is a result of our very recent evolutionary history – tracing back only some 300,000 years as a separate species – as compared to the evolution of most other mammal species around us today, which have existed for an average of at least a million years. In contrast, within a single recognised species of fruit fly in the American tropics, scientists have found more than 4 per cent genetic variation, and over 13 genetic clusters, despite all of them looking identical to one another.

The only way of finding out how much genetic variation a species contains is to sequence many individuals within each species and cover all known populations, their geographic extent and any perceived differences in form, function or behaviour. So far, this has been achieved for only a tiny fraction of species, with the vast majority having not been sequenced even once. The most time- and cost-effective way of doing this is to turn our efforts to sequencing the world's biological collections, stored in natural history museums and similar organisations like botanic gardens and universities.

When scientists began to use DNA more routinely for this purpose in the 1990s and 2000s, they were worried that historical specimens would not be suitable for genetic studies, due to their degraded DNA. Fortunately, new techniques are able to capture and assemble many fragmented pieces of DNA and have now overridden those concerns. Recently, I contributed to a study led by Colombian botanist Oscar Pérez Escobar where we sequenced DNA from basketry found in a sacred animal necropolis in Saqqara, Egypt. The object (Fig. 6), which is likely to have been used as a stopper for a jar, although we don't yet know for sure, was made about 2,100 years ago out of leaves from a date palm. Date fruits are hugely popular around the world and widely grown in Northern Africa, the Middle East and western Asia, but no one knows for sure when and where

Figure 6. Unlocking information from historical specimens. An artefact made of leaves from the date palm, found in the Saqqara necropolis of Egypt. Despite its age, it was still possible to retrieve enough DNA to estimate where this plant grew and which were its closest relatives.

people started to cultivate them. Despite the age of the specimen, we were able to retrieve thousands of fragments of DNA and compare them with sequences of current varieties. We found that the date palms that Egyptians used (*Phoenix dactylifera*) contained genes from two other wild species, the Cretan date palm (*P. theophrasti*) from Turkey and Crete (which I once saw growing by the sea), and the sugar date palm *P. sylvestris* from South Asia, which is known to be the closest relative to date palms cultivated today and contains a sap that is tapped from the trunk and consumed fresh or fermented by local people. Our study shed further light on the genetic diversity of dates and the early domestication of that valuable crop, but how those species exchanged genes, and what it meant for the size, shape and taste of their fruits, remains a mystery.

For decades, conservationists have put heavy emphasis on preserving species. Increasingly, though, they have also recognised the importance of preserving genetic diversity. Back to the question of the spacecraft posed at the beginning of this chapter – how many seeds from each plant species we should take with us to safeguard their future – the answer is *many*, or as many as it is needed to obtain a wide genetic representation of their natural diversity (something that varies considerably among species). This is why I've promoted a change of focus for Kew's Millennium Seed Bank in Wakehurst, south of London, which is the world's largest collection of seeds from wild plant species – today containing some 2.4 billion seeds from about 40,000 species. The key goal so far has been to bank as many species as possible. As Director of Science at Kew, I have ultimate responsibility for that collection, and together with my colleagues Elinor Breman and Alan Paton, we're now working with our teams and collaborators around the world to increase genetic representation *within* species of particular value, such as those with potential to help achieve food security, or that contain known medicinal or other properties of socio-economic value, and those facing the greatest risk of going extinct. For each such species, we aim to have thousands, or at least hundreds of seeds, covering the largest possible extent of their distribution, climatic tolerance and variation. This will mean the right variety can be cultivated in each region, depending on future conditions such as the amount of rain they will receive, the duration of the dry season, among other possible variations that are difficult to predict in detail. We also want to increase representation of species across all different groups of plants, capturing as much as possible of a poorly acknowledged but critical aspect of biodiversity: evolutionary diversity.

CHAPTER 3

EVOLUTION

At noon on 6 May 1930, Wilf Batty saw an unusual sight in his farm in north-west Tasmania. He spotted an animal the size and shape of a dog but with dark stripes going from side to side on its lower back. He thought the animal was trying to eat his poultry, so he rushed to grab its tail. He failed. Just as it jumped over a two-metre fence, Batty raised his double-barrelled gun and fired a shot. Proudly, he posed together with the dead animal in front of his farm shed for a photograph that would soon make the news around the world.

The animal he had shot was not a dog; it was the last known wild thylacine, also known as the Tasmanian tiger (because of its striped lower back) or the Tasmanian wolf (because of its resemblance to dogs) (Fig. 7). The farmer's actions were not unique. For decades, alongside the Tasmanian government, a farming corporation founded by a group of London merchants that planned to produce wool on the island had been offering money to anyone who would bring them the heads of the thylacine. In total, they paid out thousands of bounties, motivated by the unproven claim that thylacines attacked farm animals. The thylacine had already been exterminated in mainland Australia and New Guinea before Europeans arrived in Australia, and, besides the direct threat of hunters, it had also been struggling to survive the competition for prey from dogs introduced by Europeans to Tasmania in the early 1800s.

The species was a remarkable case of **convergent evolution**, the process by which some species come to resemble others despite belonging to distant branches on the **Tree**

Figure 7. The thylacine – one among hundreds of mammal species already driven to extinction by human activities. This drawing was based on the few video recordings and mounted specimens that remain.

of Life (also called 'evolutionary tree', 'phylogenetic tree', or simply '**phylogeny**'), which connects all living beings through branches from a common ancestor. Despite the skull of a thylacine being almost undistinguishable from that of a common dog, it was, in fact, a marsupial – much more closely related to kangaroos and wallabies. And it was remarkable among marsupials, because both sexes had a pouch.

One could argue that the thylacine was just one among thousands of mammal species, and so its disappearance from the Earth didn't matter any more than it did for other extinct species. However, what the world lost with the thylacine was not just one species, but a disproportionate loss of evolutionary diversity represented by millions of years of independent evolution. Unfortunately, the thylacine did not leave any close relatives, which would have secured the survival of the same evolutionary branch.

Our concept of evolutionary trees comes from no less than Charles Darwin, although it wasn't as eloquently articulated as his other ideas. His first 'tree' was a hand-drawn sketch in a notebook, following the simple words 'I think' (Fig. 8). An updated version later became the only figure in his seminal work *On the Origin of Species* – the foundation of the study of evolution, and probably the most important science book ever written.

Today, biologists are increasingly trying to maximise evolutionary diversity when coming up with suggestions about which areas should become priorities for conservation. This is an important transition from a long tradition

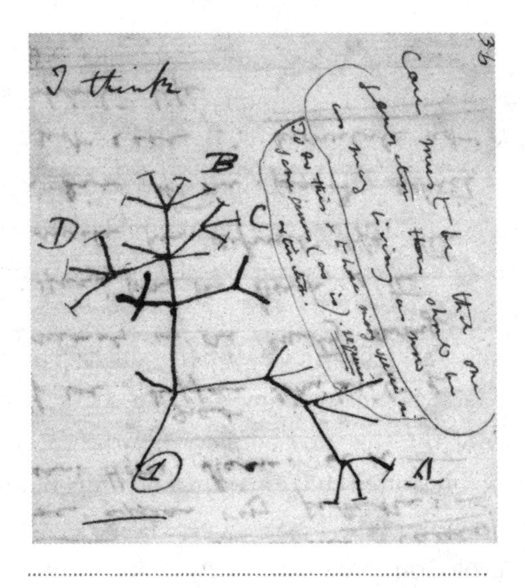

Figure 8. Charles Darwin's hand drawing of an evolutionary tree. This figure provided a succinct summary of Darwin's view on how species diverged from one another to build today's diversity of life. This method of depicting species relationships remains essentially unchanged to this day.

of counting numbers of species as if they were perfectly equivalent units. For instance, imagine there was a choice between two forested areas on which to build a new development for family homes. Suppose each of them contains two species of trees, but not the same ones. If we only cared about numbers of species (species diversity), it wouldn't make any difference which patch of forest we choose. However, species also have an evolutionary history, giving each a unique place on the Tree of Life. In this imaginary example, one area contains oaks and beeches, which have been separated by some 51 million years, when those two species last shared a common ancestor. The other area contains instead oaks and pines, which have been separated by some 313 million years of independent evolution. The second area therefore contains higher evolutionary diversity and, all else being equal, could be considered more worthy of protection.

Evolutionary diversity is often called 'phylogenetic diversity' and is measured by totalling the length of the branches of the tree that connects all the species being considered. It's often measured in time units but can also reflect differences in DNA sequences (Fig. 9).

As evolution progresses, new features emerge that help species cope with their environmental challenges and interact with other living organisms. Random mutations that make individuals more successful in producing offspring are naturally selected, and soon these mutants become dominant within a species and its descendants. Over time, this process leads to animals gaining new features, such as the lighter bones that help birds fly and the long teeth that help carnivores to capture and kill their prey. Plants end up with new chemicals in their leaves that make them poisonous to herbivores, and woody fruits and thick bark that allow them to survive in fire-prone regions. It is through evolution by natural selection that today's life forms have arrived at such a staggering diversity.

So the time has come to embrace evolutionary diversity, by establishing the relationship between species, including when and where they first evolved. This is

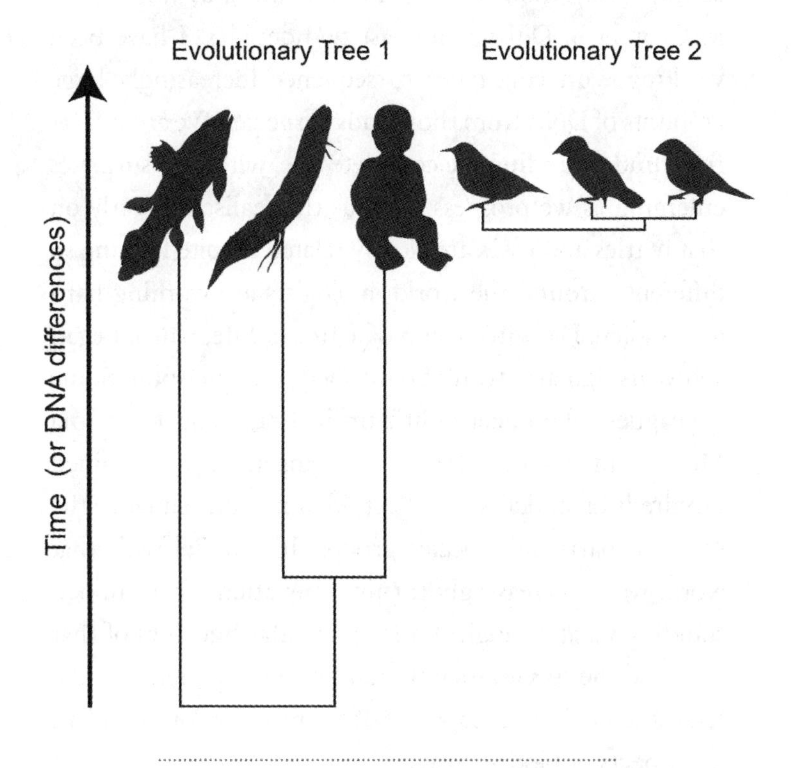

Figure 9. Examples of evolutionary trees (phylogenies). The branching patterns show how species are related, and the length of each branch is drawn in proportion to time since they diverged or the number of genetic differences. *Left:* This tree shows the coelacanth on the far left, which separated from lungfish and tetrapods some 390 million years ago. *Right:* This tree, comprising three birds in the tanager family, contains a much lower evolutionary (phylogenetic) diversity than the first, even though the species diversity is the same in both cases. The branch lengths are illustrative only.

critical for prioritising species in the conservation effort and has numerous critical real-world applications, such as improving food security and our diets, as we will see in Chapter 6. During the past two decades, I have been working with colleagues to sequence increasingly large amounts of DNA from thousands of species. We are still far from understanding the complete tree, with new surprises emerging as we progress, such as the realisation early on that nettles and roses are closely related despite looking so different. Around the world, biologists are working hard to complete Darwin's vision of a Tree of Life, which began 160 years ago and won't be finished anytime soon. Some colleagues, like Lúcia Lohmann in Brazil and Muthama Muasya in South Africa, have inspired and trained hundreds of students to collect, identify and sequence the DNA of particular species groups. In parallel with that work, we're increasingly turning our attention to understanding what ecological roles particular branches of that tree, and the species they contain, play, which remains the least studied, but perhaps one of the most important, of all the aspects of biodiversity.

CHAPTER 4

FUNCTIONS

After a year studying biology as an undergraduate, I started to become impatient. If I was to become a scientist, I knew that I needed at least three more years of university studies until I could complete a Master's degree and then one day a PhD. This seemed like too long to wait, and I needed to know if a career in science was the right thing for me.

I phoned around, looking for answers, until I heard about a project in Lapland, some 200 kilometres north of the Polar Circle, in a research station by the tiny glacial lake Latnjajaure. The station was set up in the 1960s to study the lake's food web: scientists were curious to find out what happened if they released fish into it, as they were not present naturally. As the fish didn't survive through the coldest winters, and scientific attention began shifting towards climate change in the late 1990s, the idea came to use the station to survey how a warmer climate might affect the surrounding Arctic vegetation, which was one of the ecosystems where a change might be noticed first.

The more I heard about the project, the more the words of John Muir, the famous American conservationist and explorer, resonated in my heart: 'The mountains are calling and I must go.' I volunteered to work in Lapland as an unpaid assistant for a month, if they covered my food and travel. To my surprise, despite my lack of experience, they agreed. It turned out to be one of the most rewarding professional experiences of my life, and it eliminated any shadow of doubt that science was my true passion.

Among the many tasks I carried out from early dawn to midnight sun was setting up permanent plots to study the local vegetation. We negotiated with the Sami people to fly us to various mountain summits in the tiny helicopters they would use for herding reindeer across the vast, open landscape. On those mountains, we hammered down aluminium sticks, which could withstand wind and snow, at precise locations. We then recorded every species of plant and lichen within each plot. This would allow researchers to come back at regular intervals to survey how biodiversity might change over time. The initiative, launched simultaneously at 18 European locations during that summer of 2001, soon grew to over 120 sites around the world, from the arctic and temperate regions in both hemispheres to the tropics.

It was difficult to predict what would happen. The scientists I talked to were expecting several Arctic species to disappear soon from the local landscape as it warmed, reducing the diversity of the flora. To everyone's surprise, however, as the temperature increased, subsequent surveys showed instead an overall *increase* in species richness over just 10–15 years. What changed even more dramatically was the composition of species in these plots: some became far more common than others, some moved elsewhere, and some appeared from lower altitudes or other mountains far away. Because each species differed in their ecological **function** – the ways in which each of them affected their surrounding environment, and interacted with other species – the most important insights those recurrent surveys of vegetation revealed were striking changes in the plots' *functional diversity*, the total variety of ecological functions those species perform. Those differences included how much total weight the different species accumulated and whether they could survive for many years or just one season. Plant communities began to store more carbon than previously, and shrubs started to overshadow and replace herbs, mosses and lichens. In general, species with large geographic distributions – often spanning multiple countries or even continents – became more common, making the plant communities more homogeneous, or similar to each other. What that research showed was that climate change was leading to visible and substantial changes in the functions of plant communities over extremely short periods of time, threatening the identity of high-elevation habitats and their specialised species.

Functional diversity is the fourth axis of biodiversity that is intrinsic to the living species on Earth. Crucially, compared to the three other biodiversity variables I've described, it's often the first variable to change as the environment changes. As in the previous chapter, if we were to select two hypothetical areas for conservation where all else is equal, and we have the same number of species, genetic diversity and evolutionary history, the site with the greatest biodiversity is the one in which species are performing the largest total number of functions. Unfortunately, as climate changes and certain species take over large areas and outcompete others, they often bring a narrower set of functions than the original species – such as a dominance of fast-growing plants that overshadow others, or generalist animals that feed on many other species. As such, functional diversity is critical to maintaining a natural world full of variety and difference.

Functional characteristics of biodiversity, often called traits, are measurable features that allow species to fill diverse roles. They can relate to physical features (such as the size and type of plant leaves), behaviour (such as the diet of animals: carnivorous, herbivorous etc.) or habitat (such as species restricted to deep or shallow water). As you can see, functional traits include very different things and can be highly organism-specific. In practice, it's nearly impossible to identify and measure every function a species plays, and it can be difficult to measure even some of the most important ones.

As an example, in recent years there has been massive interest in growing forests around the world to slow down climate change, either through active tree planting or natural regeneration. Thoroughout the growing season, trees capture large amounts of carbon dioxide – one of the most important greenhouse gases – through the tiny pores in their leaves, called stomata. Using solar energy, the tiny green organelles inside the leaf cells – the chloroplasts – are then able to transform carbon dioxide and water (transported all the way up from the roots) into sugar and oxygen. I am referring to photosynthesis, the single most important chemical reaction on Earth, which directly or indirectly sustains the existence of nearly all species. The sugars created are used as the backbone molecules to almost everything we see in a forest: the wood, leaves, roots and fruits.

The potential of trees to help capture and lock down carbon from the atmosphere is, however, quite uncertain. The issue is not only that we lack comprehensive assessments of carbon content in the wood of the world's 70,000+ known tree species; we also have little idea of how much carbon trees can store *under* the ground in their roots. To date, scientists have dug up very few trees to measure the total weight of their root system and carbon storage, but the trees that have been measured have differed by an order of magnitude depending on the species. Without actual measurements to hand, any calculation of a forest's total carbon storage can differ a great deal from its reality. If the truth were known, it could significantly affect our conservation priorities where climate mitigation is the main goal.

An interesting aspect of functional diversity is the concept of **redundancy**, where species play similar functions to others, and therefore there is a surplus of a function in an ecosystem. Is this a good or bad thing? A couple of years ago, I contributed to a study led by Colombian palaeobiologist Catalina Pimiento, in which we analysed thousands of mollusc fossils from the Caribbean. I was fortunate enough to work as a dive-master there for several months, and as anyone who has snorkelled there will know, there is staggering diversity in the marine life of the region. About three million years ago, however, some still unknown event (presumably environmental change) led to a massive loss of species. In this study, we were curious to find out what allowed some species to survive while others didn't. We discovered that 'redundant' species, such as those feeding on tiny floating particles and organisms (Fig. 10), were disproportionally impacted. Although such redundancy was bad news for

Figure 10. Species of a rocky shore. Healthy sea communities perform several different functions that ensure nutrients are recycled, each of them by several different species. Suspension feeders, for instance – like mussels, sponges and barnacles – eat plankton (small free-floating organisms, or the eggs and larvae of larger animals) and other organic matter in the water. In turn sea stars, crabs and fish – all of them predators – eat the suspension feeders. Meanwhile, diverse brown, green and red algae absorb and store nitrogen and carbon from the water. It is this abundance of functions, composed of various species and complex food webs, that makes healthy ecosystems resilient to climate change, human disturbance or disease: if one species disappears, others can take over and keep the system stable.

Predators
Suspension feeders
Nitrogen and carbon fixation

those species individually (which did not survive as well as more specialised species, perhaps because they faced higher levels of competition for food), it was good news for the Caribbean ecosystems as a whole, which only lost 3 per cent of all ecological functions – despite losing half of the number of different species.

Our study, along with others, has shown that redundant species provide an extra layer of insurance: even if some of them go extinct, the overall functional diversity remains. When functionally *unique* species go extinct, however, the ecosystem suffers a much bigger loss. So along with the other axes of biodiversity, understanding and measuring functional diversity is crucial if we are to fully understand and protect biodiversity. It helps us to value and prioritise natural habitats for biodiversity protection and carbon storage. Finally, functional diversity provides a bridge between the smallest components of biodiversity – species and their genes – to the largest of them: ecosystems.

CHAPTER 5

ECOSYSTEMS

In 1802, after months travelling through the mosquito-infested jungles of South America, the great German naturalist Alexander von Humboldt visited the Chimborazo volcano in Ecuador and scientifically documented for the first time how the species of plants and animals changed the higher he climbed. He saw how climate and topography helps to create great variation in ecosystems – from the hot and humid lowlands, through the misty and densely packed cloud forests, to the cool grasslands and snowbed vegetation of the Andes. Through his travels to the Caribbean, North America and Asia, and in extensive correspondence with other scientists, he described how a similar diversity of ecosystems occurred around the world, each with its own set of species and characteristics – from the fire-prone savannahs of Africa to the seasonal meadows of Canada and the alpine grasslands of the Tibetan Plateau. In the oceans, temperature and topography interact instead with salinity and depth to produce such habitats as coral reefs, seagrass meadows, mudflats and the deep-sea floor, where life might have begun some four billion years ago in hydrothermal vents.

The diversity of ecosystems on Earth is a key reason we have so many species, and why they exhibit such large variation in life forms, behaviour and functions. In Madagascar, for example, the succulent woodlands of the south-west contain a totally different set of species to the lowland rainforests of the east. In many cases, focusing on the conservation of whole ecosystems is more effective than the traditional focus on single iconic species. This is because species can rarely be protected in isolation, and saving a portion of an entire ecosystem will help protect many species simultaneously.

In nature, species are made up of populations, and populations consist of individuals. At higher levels of organisation, several species group into communities (such as the rocky coast community described in the previous chapter, Fig. 10), and communities group into increasingly larger units. A common terrestrial scheme, adopted by the World Wide Fund for Nature, has been to aggregate terrestrial ecosystems* into eight realms (most of them famously identified in the 1800s by naturalists including British explorer and co-founder of the theory of evolution, Alfred Russel

---

* In contrast to the principles of **taxonomy** – including Linnaeus's classification of species into genera, families, orders, classes and kingdoms mentioned previously – the terminology, identification and delimitation of marine and terrestrial ecosystems remain diffuse among scientists, and terms such as ecosystems, biomes and bioregions are sometimes used interchangeably. Here, I use the commonest of those terms: ecosystem.

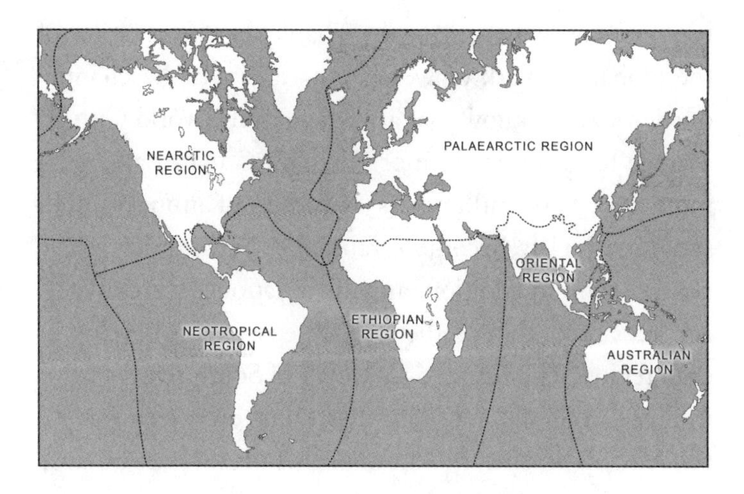

Figure 11. Wallace's Zoogeographic regions. Through his own extensive travels and studies of the literature available at the time, Alfred Russel Wallace was able to propose a division of the world's major faunistic (animal) realms in 1876 that remains mostly valid to this date. Each of these regions contains animal groups peculiar to that region. One example is the animal order Xenarthra, which includes anteaters, sloths and armadillos – animals that are almost exclusively found in the Neotropical region, where they once evolved in isolation from other continents.

Wallace), such as the Neotropical and the Palaearctic regions (Fig. 11). The realms between them contain 14 biomes, such as boreal forests and mangroves, and these are further subdivided into 867 ecoregions, such as the Temperate broadleaf and mixed forests. In general, the finer the scale, the more useful a designation becomes for real-world conservation.

It's fascinating to note how well Humboldt's early observations on the connection between climate and type of ecosystem still apply. Thanks to the creation of the global network of weather stations that he championed, today we know that anywhere in the world with an average temperature of more than 18°C (64°F) and with more than 2,500 millimetres (98 inches) of annual rainfall should naturally become a rainforest (well, unless we've cut it down). In addition, annual variation – seasonality – is a key aspect that predicts which ecosystems develop in each region. The Cerrado savannah of South America, for instance, receives more rain than some evergreen forests, but remains an open mosaic of grassland and trees because the rain is restricted to just a few months every year. For a tropical rainforest to form, it needs to receive at least 60 millimetres (2 inches) of rain every month.

Human disturbance also plays a role, since we have removed many of the large animals that kept forests open and counterbalanced that by introducing fire and cattle, meaning that our actions have altered the pre-human dynamics of many terrestrial ecosystems. Most recently, we have done the opposite in many places – we now prevent and rapidly try to extinguish forest and shrubland fires, even those caused by natural phenomena such as lightning. This is most problematic in the world's Mediterranean zones: the Mediterranean Basin, California, Central Chile, Southwestern and South Australia and the Cape Region of South Africa (coincidently, the regions known to produce the best red wine). But it is also an issue in other regions, where fire is not as regular, but does happen occasionally, such as Scandinavia, northern Asia and parts of Canada.

It may seem counter-intuitive to some, but fire prevention often brings negative consequences for the ecosystems and species that are adapted to – and sometimes dependent on – fire, for successful reproduction or growth. Another side effect of regularly extinguishing fires is that old plant material (such as leaves and branches) eventually accumulate in great quantities, making the fires much hotter, larger and more devastating than under more natural conditions.

Sometimes the transition between ecosystems is so abrupt that you can drive over a mountain pass and go from a semi-arid scrubland into a dense forest in the space of just a few metres. This is a ubiquitous phenomenon seen in many mountains around the world, like on the island of Gran Canaria and in the Cape region of South Africa, where moisture-carrying air currents foster the maintenance of humid vegetation on one side of the mountain (the windward side) and dry conditions on the opposite (leeward) side. In other places, the transition is gradual, taking place over thousands of kilometres, such as between the semi-arid cacti-filled landscapes of northern and central Mexico, which turn into dense, moist forests in southern Mexico and Central America.

Ecosystems, like species, are transient. Some major ecosystems we are familiar with originated very recently in Earth's history, while others no longer exist. Global climate changes can lead to changes across continents, such as the seemingly simultaneous appearance of vast grasslands in Africa, South America and Australia. Peruvian botanist Mónica Arakaki, together with American evolutionary biologist Erika Edwards and their colleagues have shown that those ecosystems expanded as a consequence of a cooling and drying period beginning some 13 million years ago, which triggered the diversification of grasses and succulents (water-storing plants) performing new kinds of photosynthesis, particularly effective in dry environments, called C4 and CAM (Crassulacean acid metabolism). Compared with C3 photosynthesis, characteristic of the vast majority of plants in the world, C4 plants – such as many savanna grasses – concentrate carbon in specialised cells around the enzyme that produces sugars, without contact with oxygen or the need for respiration (a process that loses a lot of water to the atmosphere). In contrast, CAM plants, like cacti, save water by opening their stomata at night, when temperatures are cooler and they can take in and store carbon dioxide in chambers for later assimilation during the day. Perhaps counter-intuitively, it was those miniscule adaptations at the cellular and molecular levels which enabled the turnover of entire ecosystems.

Conversely, fossils reveal the loss of a massive ecosystem – the boreotropical forest – which emerged soon after the demise of dinosaurs around 66 million years ago. Although you have probably never heard of that forest, it was the dominant vegetation covering a huge area in the southern parts of North America, Europe and Asia for over 20 million years. It contained an unusual mixture of large trees of tropical origin, but with thick, drought-adapted leaves. We don't know details of the climate at that time, but, based on those characteristics, it's reasonable to assume that the region had clear wet and dry seasons. It was probably drier than today's rainforests, and wetter than today's regions that have a Mediterranean climate.

We know little about the evolution of ecosystems, however, and what this means for long-term conservation. These historical changes tell us that we shouldn't take today's ecosystems for granted. A particular concern is the likely existence of 'tipping points', or 'points of no return', beyond which a whole ecosystem may not be able to revert into a previous state. Scientists estimate, for instance, that the world's largest rainforest, Amazonia, could turn irreversibly into a savannah from the point at which 20–25 per cent of its total forest is lost. As we've already lost some 18 per cent of Amazonia's pre-human forest area, according to some estimates, this could mean we are getting very close to that point. We've seen similar shifts in ecosystems in recent times, such as the desertification of the Sahel region which probably happened by natural causes. The large belt of arid land south of the Sahara covers an area three times the size of Egypt and was wet and largely forested until just a few thousand years ago. Even more dramatic was the collapse of the Aral Sea, a large body of water between today's Kazakhstan and Uzbekistan that teemed with life and around which humans built a rich social culture. It all disappeared, due to the decision by the leader of the Soviet Union, Joseph Stalin, to set up large plantations of irrigation-demanding cotton which was a totally unsuitable crop for the region and destroyed its wildlife.

Ecosystems are the largest component of Earth's biodiversity. Although they are immensely larger than genes – the smallest unit of our 'five-pointed biodiversity star'– both genes and ecosystems represent two lenses of the same telescope. All five biodiversity 'lenses' of that telescope are aligned and complementary, and just like a fine-tuned telescope allows us to explore distant celestial bodies, so the biodiversity telescope allows us to explore, understand and really see our living planet. Now that we understand more fully what biodiversity is, it's time to understand why it really matters, before we face the hard facts of why it is disappearing so fast.

PART TWO

THE VALUES OF BIODIVERSITY

How much is a flower worth? It really depends on who's answering. In our excessively monetised societies, the economist may attempt – more or less objectively – to put a price tag on it. Flowers of timber trees, essential for their reproduction and therefore their very existence in a forest, may have a value that is the price of the timber itself. For the smallholder, the value will only be high if the flower belongs to a crop that survives this year's drought. For a bee that pollinates a single species of plant and whose tongue cannot reach the nectaries of any other, finding the right flower to satiate its hunger will be a question of life and death. For the poet or the nature lover, the value is at least as high as the long hike in the forest in search of rare blossoms that enchant our lives and fill us with awe. Values are relative, fluid, critical yet diffuse. And as stars care little about what we think about them when we gaze into the void, flowers – and biodiversity – are not really here for us. But one thing is certain; we cannot be here without them. And here is why.

CHAPTER 6

FOR US

I remember very clearly the day in the early 2000s when I knocked on the door of Lennart Andersson in the Institute of Botany at the University of Gothenburg. It was a hot and damp day, shortly before the summer break. I had got to know Lennart as a quiet, reserved and rather eccentric lecturer who spoke so softly that we students always needed to sit right at the front of the classroom to have a chance of hearing him. I didn't realise he was a professor and, in fact, a famous scientist – internationally renowned as a leading specialist in the flora of the American tropics, and for having scientifically named over 150 plants – until many months after I first met him. He didn't like to blow his own trumpet.

Lennart was sat in his sunken old chair, typing frenetically on a dirty keyboard with only his index fingers. His desk was filled with tall piles of printed articles, journals, maps, herbarium specimens and several empty coffee cups. I cleared my throat and excused myself for interrupting him. I said that I had been thinking about my future, and wanted to know whether he might be willing to supervise my science project for my Master's degree. It took him a few seconds to digest what I had said, before a large smile spread across his face. 'Sure,' he said and pointed to the chair on the other side of his table. 'Let's talk.'

That day changed my life. For hours, we engaged in a deep discussion, which soon turned to a group of South American plants collectively known by the scientific name Cinchoneae – a group of some 120 species of small trees. The group is best known as the source of quinine, a bitter substance present in the bark, which is now famous for giving its taste to tonic but had also been the only known treatment for malaria for hundreds of years. Some think that quinine is the plant drug that has saved the most human lives in history. Lennart had spent many years trying to figure out just how many species the group contained, and how to tell them apart, but he lacked specimens from a key region: the north-western part of the mighty Amazonian rainforest. Would I be interested in working with that group of plants, and perhaps even travelling there with him? Well, you could guess my answer. The rest is history: travels in a tiny airplane and canoes through the forest; long hikes in the heat; many collected specimens; and the scientific description of a new plant genus containing two species. Although some fellow botanists choose to name new plants after their partners or relatives, I decided to call the genus *Ciliosemina,* meaning 'hairy seeds' – something that was easy to see with the naked eye (Fig. 12), and would help people distinguish them from other species in that group, which all had glabrous (hair-free) seeds.

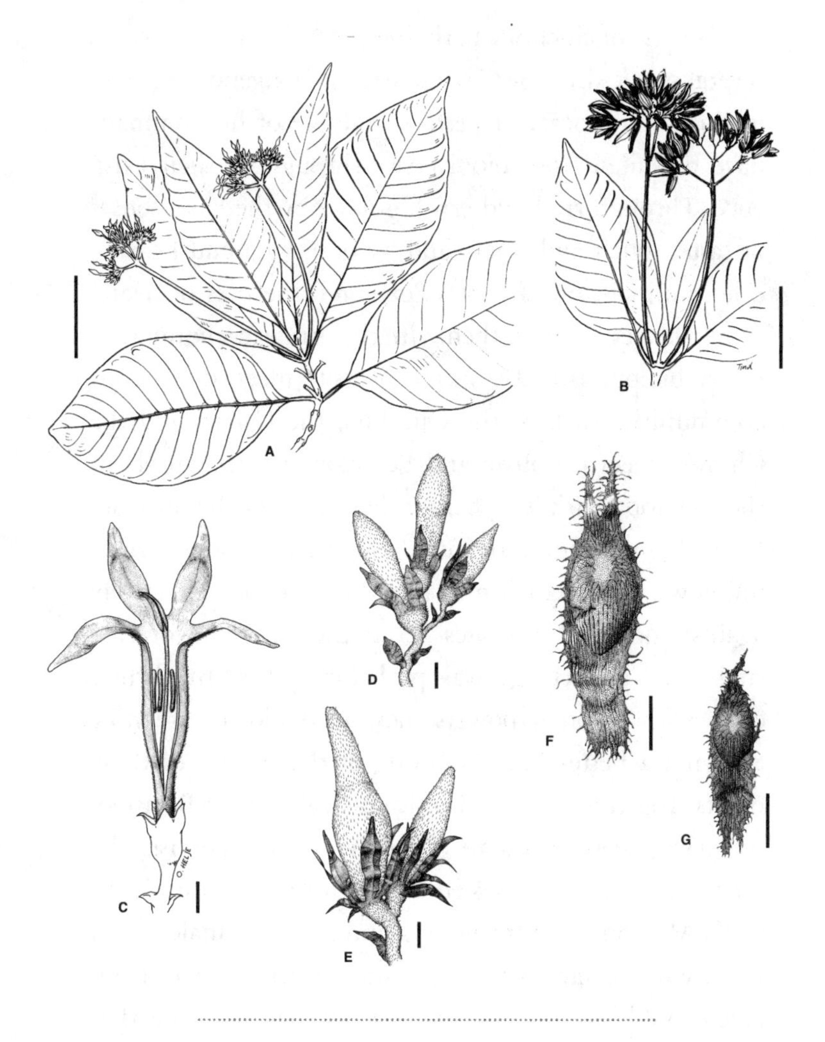

Figure 12. Two species in the plant genus *Ciliosemina*. I described these in one of my first scientific publications. The group to which they belong – the coffee family, Rubiaceae – is one of the most diverse plant families in the world, every year yielding new scientific discoveries, primarily from understudied tropical regions.

The use of cinchona bark (the general term ascribed to several species in tribe Cinchoneae) by indigenous peoples in South America is a perfect example of how humans have benefited from biodiversity throughout their existence. Through trial and error, guided by the basic senses of taste, touch and smell, and by observing and hearing other animals, our ancestors explored the uses of nearly every species around them. In the Andean mountains, where biological and cultural diversity meet, indigenous communities such as the Quechua, the Cañari and the Chimú in Peru, Bolivia and Ecuador had known about the cinchona bark long before the arrival of the Spanish. How those communities used the bark remains largely unknown, although it may have been applied effectively against intestinal parasites. Even though some of that traditional knowledge was probably lost in the brutal clash with colonial powers, new technologies are now allowing a better understanding of the history of these plants. Together with my Danish colleague Nina Rønsted, we have been fortunate enough to co-supervise the doctoral theses of two excellent South American students, Carla Maldonado from Bolivia and Nataly Canales from Peru, who generated vast amounts of genetic data from recent and historical collections of bark samples. Together with a third student Kim Walker and Kew's Economic Botany Collection curator Mark Nesbitt, their studies have elucidated many aspects of this intriguing saga. And the cinchona bark is by no means an isolated example. To date, scientists have been able to compile data from some

40,000 plants of known uses by traditional communities around the world, as sources of medicine, food, fibre, shelter, timber, poisons, energy, oils, ornamentals, narcotics, and much more.

The multiple facets of biodiversity that we have explored in previous chapters also play an important role in our exploration and use of species. We often hear how important it is to have a varied diet, rich in fruits and vegetables. But which foods should we choose? This is obviously a complex question, which involves aspects like season and availability, environmental impact, taste and price. One additional factor that has been largely neglected is their evolutionary diversity which, as we have seen, can vary a great deal among species. If you prepare a dish with potatoes, tomatoes and aubergine, you'll have three members of the same plant family (Solanaceae) – all closely related, representing some 37 million years of evolution. If instead you choose potatoes, broccoli and walnuts (from the Solanaceae, Brassicaceae and Juglandaceae families), you'll increase this by almost tenfold, to 340 million years.

The crux here is not how much evolutionary time you get on your plate, but the nutritional consequences of that elapsed time. We can get our daily macronutrients – those that provide the bulk of our energy intake – from many different food sources. These are the proteins, fats and carbohydrates we need. Besides those, we also need a whole series of micronutrients, also called vitamins and minerals, which we cannot synthesise in our bodies but which are essential for our survival. Those are especially hard to find and are often constrained to particular branches on the Tree of Life, rather than being randomly scattered. Take zinc, for instance. We need it in tiny amounts – some 8 milligrams a day for an adult woman – but it is used in every cell of our body, as it is key to making DNA. It also helps us fight off bacteria and viruses. Some of the richest plants in zinc all belong to the legume family (Fabaceae), including chickpeas, lentils and kidney beans. For another micronutrient, selenium (important for our fertility), you'd want instead to eat members of the crucifer family (Brassicaceae), which includes cabbage, cauliflower and broccoli.

Although more research is needed to prove this case, logic and available evidence suggest that an evolutionarily diverse diet is good for us. It's very worrying, therefore, to consider the fact that this is often not the case. Worldwide, over four billion people rely on three staple foods: rice, maize or wheat. As you'll note, these are all grasses (Poaceae) – meaning that our diet is essentially very similar to that of a free-range cow. Over 90 per cent of the human calorie intake comes from just 15 crop plants. These numbers stand in stark contrast to the 7,000-plus plant species that scientists have documented as edible, based on traditional knowledge worldwide. And this number still only represents a fraction of the total edible species that exist across the world.

Our societies' dependency on so few crops is clearly contributing to malnutrition, poverty and inequality. In addition, we place ourselves in an incredibly vulnerable position by relying on so few species, since a single pest or pathogen could rapidly wipe out vast plantations. This is precisely what happened in Ireland during the 1845–1849 Great Famine, when a fungus-like organism destroyed the potato crop, the dominant food source for much of the population at the time. The failure of the crop, combined with other socio-political factors stemming from inequalities experienced by large swathes of the native population under colonial rule, contributed to the tragic deaths of millions of people. Today, another fungus is posing an as-yet untreatable threat to the world's most consumed fruit: the banana. Despite the fact that over 1,000 varieties exist – each with its own genetic variation, colour, shape and size – half of the world's banana production and 99 per cent of exports rely on a single variety: the Cavendish. The latest and most devastating fungus strain was first reported in East Asia around 1990, and has since then been found in Australia, Africa, Middle Asia, and in 2019 also Latin America, making it a pantropical disease. The blight is likely to affect most, if not all, monocultures of Cavendish banana, so those farmers who grow a higher diversity of varieties are much better off. In Africa, smallholder farmers have grown and traded multiple banana varieties for over 1,000 years, contributing to three-quarters of the continent's total banana production, and safeguarding their produce from total wipe-out.

The value of biodiverse ecosystems extends far beyond that of individual species, making them much more than the sum of their parts. For millennia, they have provided people with numerous tangible contributions. These are often called ecosystem services or, using a wider and more recent concept that more clearly encompasses non-material benefits, **nature's contributions to people**. Some of the most beautiful and biodiverse forests I've visited around the world have been protected because they provide a source of clean water to cities. Wild bees and other insects pollinate many of our crops for free. Forests and other ecosystems, including natural parks, give us beauty, fresh air and opportunities to exercise, helping us and our families to recharge from stressful lives, improving life quality and wellbeing.

Although plants are a primary source of the benefits we have derived from nature, we have also taken advantage of numerous animals. We have obviously used them as food, and they provide us with valuable proteins and fat, often in intricate ways – from the traditional seal dishes of Greenland's Inuit to the edible birds' nests created by swiftlets using solidified saliva in Borneo's caves. But we've also used animals in medicine, such as horseshoe crabs, native to the Atlantic coast of North America, whose bright blue blood plays an invaluable role in medicine due to its extreme sensitivity to toxic bacteria. Horseshoe crabs have been around for more than 240 million years, and even with today's advances in medicine, the use of their blood is still the most effective way of ensuring medicines and vaccines are free from harmful bacteria. Each vaccine against COVID-19 produced in the USA to date was tested in this way to determine whether it could be licensed for use.

Despite so many reported uses of biodiversity, we've only scratched the tip of the iceberg – there is an immense treasure trove of useful properties to be unlocked. We have no idea what the next pandemic might be, but its cure might well be hiding in the forests of Congo or the grasslands of New Zealand. Every species alive today carries genes that have evolved through millions of years to enable them to cope with specific environmental conditions, fight off viruses and bacteria, and develop clever innovations to outperform other species. The unpleasant blue or green mould that grows on stale bread was despised for centuries until Scottish scientist Alexander Fleming discovered, quite by accident, that it produced a substance – penicillin – that would go on to save hundreds of millions of lives. That was but one species of fungus among at least three million that are estimated to exist, so there are likely to be many more useful fungi awaiting discovery. Today, new technologies can speed up the detection and testing of important properties in different species, guided by our growing understanding of how species relate to each other, and what roles their genes play.

In economic terms, biodiversity is often considered an 'asset'. Similar to economic portfolios, the more biologically diverse options you have, the better your chances of persisting when faced with adverse conditions, such as extreme climate events and unexpected environmental threats. Small farmers have long understood the value of diversifying their crops. In Eastern Africa, for thousands of years people have been cultivating a wide range of species and varieties on their land. They have chosen crops that are particularly adapted to their local environments, from the floodplains to the highlands. Such variety has given them long-term security, for example if one particular crop didn't do so well in a particular year, as a consequence of an usually long dry season, or due to the attacks of a locust swarm, then others may not be affected.

The major challenge of deriving value from biodiversity boils down to a single word: sustainability. Throughout our history, we've taken for granted that we could take whatever we needed from nature, as much as we wanted, without giving back or allowing it enough time to recover. We've caught too many fish, killed too many horseshoe crabs, taken down too many quinine trees, hunted too many seals, stolen too many edible birds' nests. We have thought of biodiversity as an infinite asset, stripping it of its genetic, taxonomic, evolutionary and ecosystem diversity. Our attempts to restore ecosystems are commendable, but risk falling short unless we also stop the unsustainable harvesting that is fuelling their deterioration – a major driver of biodiversity loss, as we'll examine closer in Chapter 10. We must urgently find solutions for sustainable social and economic development, in ways that are far more equitable and environmentally sustainable. We also need to recognise that species are not a collection of useful items awaiting our use: they are complex, intertwined organisms playing vital roles in the maintenance of well-functioning ecosystems, which are the prerequisite of a healthy natural world.

CHAPTER 7

FOR NATURE

The view that biodiversity primarily exists as a resource for humans is deeply rooted in our culture and many religions. In the Bible, the book of Genesis reads, '*I have given you every herb bearing seed which is upon the face of all the earth, and every tree in which is the fruit of a tree yielding seed*' and says that humans '*have dominion over the fish of the sea, and over the birds of the air, and over the cattle, and over all the wild animals of the earth, and over every creeping thing that creeps upon the earth*'.

The exploitative view of biodiversity still dominates most people's minds and public debate. It is also the most widely used argument for conservation: that we must protect species because they may hold some known, or yet unknown, benefits to us. During my many trips collecting plants around the world, I can't keep count of how many times I've been asked by the local people, 'What's that plant good for?' followed by a sign of disbelief if I replied, 'Nothing that I know of.' In Sweden, the common tick – one of the country's most insignificant yet dangerous animals, given the diseases it can carry – is a target of radio debates almost every summer which ask why on earth they exist, and how one might go about killing them off once and for all.

But every species is part of nature's intricate web of life. Their existence is crucial to the health and function of ecosystems, as they drive key natural processes, such as feeding, reproduction, dispersal, competition, survival and mortality. Some species may play roles we are barely aware of, such as the chemical warfare taking place between different fungi inside a rotting log, or the shrimps, sea snails and bristle worms that can reduce the carcass of a large whale at the bottom of the sea to a pile of bare bones. Even the Swedes' most hated ticks play key ecological roles: they are food to many animals, such as birds, frogs, toads, lizards and spiders; they carry viruses, bacteria and microorganisms between many animals, probably helping to regulate their population sizes; and they host several parasitic species, including the tick wasp, which depends on them to lay their eggs and survive.

At times, the impact that a single species can have on an ecosystem is anything but subtle. In 1995, biologists released eight wolves into America's Yellowstone National Park. It had been seven decades since the species was locally exterminated due to hunting by local farmers with the support of the public authorities, since they would occasionally prey on cattle. Already in the 1930s, scientists were concerned by the impact of the growing populations of a species of deer called the Rocky Mountain elk. The deer, which had been previously hunted by the wolves, were now intensively grazing in the park, increasing erosion and putting many plants at risk of disappearing. Carrying out such a 'rewilding' experiment wasn't without controversy, as many people were nervous that the wolves would leave the park's boundaries and attack the livestock of farmers, or pose a threat to people.

In the years that followed, the park rangers and biologists involved in the project couldn't believe their eyes. The eight wolves that were released into the park set in motion a cascading chain of effects of far greater impact than anyone could have imagined, and which is still unfolding. As predicted, the wolves did bring down the deer population. With fewer deer, the park's valleys soon started recovering from overgrazing and the vegetation increased. Many plant species increased in numbers, including aspen, cottonwood, alder and several species of willow and berry-producing shrubs. In particular, willows were a main source of food for deer during winter, but also for beavers. With more willows, the single beaver colony that had survived in the park until the wolves were reintroduced now had plenty of food to thrive and multiply. As they did so, they started affecting the whole park's hydrology, by spreading and building new dams and ponds. Those water bodies in turn provided suitable habitats for many fish and other freshwater species. Above ground, the bird life thrived in this newly mosaicked landscape and the number of songbirds increased. In other words, the introduction of a single species changed the dynamics and biodiversity of an entire ecosystem, even changing the course of its rivers (Fig. 13). And, in contrast, the feared impacts on livestock and people have been minimal and manageable.

Species that have such critical effects on nature are often termed **keystone species**. Other examples include elephants, which eat and trample on vegetation, preventing the tree canopy of forests becoming too dense and facilitating the passage of light down to the low-growing species. Sea otters control the population of sea urchins by eating them, which helps to maintain the balance of the highly diverse and intertwined kelp forests off California's coast. Many woodpeckers create new nests in tree trunks each year, and old nests provide shelter for many other species of wildlife, from owls, ducks and swallows to many small mammals.

Since no species lives in isolation, removing one has a direct effect on another. During my PhD, I studied a group of plants in the bluebell family, the lobelioids. I discovered that they originated in southern Africa, and from there were able to colonise landmasses thousands of kilometres away, probably thanks to their tiny seeds that could either have been blown across this vast distance (I figured out that a single gram contains some 36,000 seeds), or hitched a ride after getting stuck in bird's feathers or feet. Whichever mechanism was responsible for this epic journey, at least one seed landed in the Hawaiian archipelago many millions of years ago, which thrived and gradually gave rise to an extraordinary group of over 125

Figure 13. Yellowstone National Park after the re-introduction of wolves. The few individuals released into the wild led to a series of events that created a diverse, heterogeneous ecosystem very different to the one that had been formed after wolves went locally extinct in the 1920s.

species found nowhere else on Earth – the largest burst of speciation of plants on any archipelago.

One of the reasons why the Hawaiian lobelioids diversified into so many species was because they developed adaptations to the local fauna, in particular birds. Over time, and as a result of natural selection, several passerine birds evolved beaks that perfectly matched the shape of specific species of lobelioid, while the flowers evolved perfectly shaped petals to aid the birds as they fed on their nectar. This created a mutually beneficial relationship where the birds received an efficient source of food, while the plants got an effective pollinator, which would fly long distances to find a flower of the same plant species to feed and simultaneously deposit pollen. This beautiful co-evolutionary interplay continued for millions of years, until Polynesians arrived more than 1,000 years ago, bringing with them pigs and rats; and then later the Europeans landed in the archipelago, importing an even more ferocious animal – the domestic cat. Abandoned cats soon became feral, surviving by preying on the local fauna and driving several endemic bird species to extinction, including those that interacted with the lobelioids. Without their optimal pollinator, many lobelioid species dropped massively in numbers, and several are thought to have gone extinct.

Besides supporting the intricate and fragile interactions among individual species, biodiversity underlies the resilience of ecosystems to natural and human-driven disturbances, from hurricanes to bulldozers. This is because high species richness is often able to maintain a high level of functional diversity: if one species drops out, at least temporarily, another can replace it. Most monkeys, for instance, aren't too picky when it comes to food sources. In one of my visits to Central America, I learned about the work of researchers who had mounted GPS tags on the monkeys of Barro Colorado Island in the Panama Canal. The tags showed that those animals will browse large tracts of forest every day searching for food, travelling long distances. As long as one tree species carries fruits, or enough insects or small prey are around to catch, the monkeys' needs will be satisfied, and so their existence and all their contributions to the ecosystem will remain.

The importance of biodiversity for nature, from the tropics to the polar zones, applies regardless of how its ecosystems are naturally diverse or poor in species. In the Arctic, polar bears – weighing up to 800 kg – feed almost exclusively on seals, and occasionally walruses, bird eggs and whale carcasses. Any changes in the population sizes of polar bears or seals as a result of climate change or hunting could therefore lead to immediate disruptions of the food chain, with knock-on effects across the Arctic ecosystem.

Although I've separated out the value of biodiversity for people and for nature into two different chapters, it's fair to say that in most cases, what is good for nature is also good for us. A thriving mangrove will not only provide the necessary habitat for many forms of marine life, it will also give people food and protection against storms and tsunamis. A well-protected rainforest will not only support all axes of biodiversity, but it will also give people a wealth of invaluable ecosystem services to rely on. This is all about functions and benefits – but what if a species has not a single known or presumed contribution to others? Would we still be able to justify investing resources to support its continued existence?

CHAPTER 8

FOR ITSELF

In 2017, the *Washington Post* published an open editorial letter by a scientist, under the title, 'We don't need to save endangered species. Extinction is part of evolution.' Although the headline of the scientist's letter was chosen by the newspaper's editors, as is common practice, the article contained many sentences that perfectly reflected its message, such as: 'The only reason we should conserve biodiversity is for ourselves'; 'Conserving a species we have helped to kill off [...] serves to discharge our own guilt, but little else'; and 'extinction does not carry moral significance'.

Seeing those views published in a leading newspaper with over 100 million monthly readers got me really upset. I felt that calling the extinction of current species a 'natural' process (it is not: species are now disappearing hundreds to thousands of times faster than during pre-human times), and using that assertion as a reason for de-prioritising nature conservation, was both misconstruing and excessively anthropocentric. Scientists do occasionally make fools of themselves by speaking out on contentious topics far from their comfort zone, and although their facts might not be straight, I applaud their wide societal engagement. In this case, however, the views were coming from a well-regarded, young researcher whose expertise was precisely in the subject to hand: biodiversity science.

While many comments from angry people immediately started to appear on the newspaper's website, on social media and in blogs, I wanted to show that the views expressed in the editorial were not broadly shared by biodiversity scientists. So my friend and colleague Allison Perrigo and I drafted a response to the article, and started reaching out to a handful of other colleagues to ask if they would be interested in co-signing it. Then things started to move really fast, and almost uncontrollably. By word of mouth, our initiative started to gather dozens, then hundreds, and finally over three thousand other signatories, including Nobel prize-winners and many other prominent figures in science and society. The response was finally published by the *Washington Post* after a lot of push from us. Although it was much shorter than the original controversial article due to a stringent word count – and was probably read by just a fraction of its readers – it did show just how passionate many people feel about the imperative to protect species, not only for their human and ecosystem values, but for their own intrinsic value.

It strikes me that this whole discussion probably would never have happened if, instead of other species, we were talking about human beings. Of course, the vast majority of people around the world don't have any links to any of us; they are too far away to be part of our local community, and they are neither growing our food nor receiving our services. They may matter to their relatives and friends, but many people don't have any immediate descendants or close family members, and this doesn't make them any less worthy of society's care, support and appreciation. In other words, every person has their own value and their own right to survive and thrive. Shouldn't this idea extend to all living things, too? I know I'm treading on sensitive ground, as many people would argue there is an ocean's distance between humans and all other species. But, as an evolutionary biologist, I see the overwhelming evidence that we are but one among many other species of great apes and monkeys; a single leaf on the primate branch of the Tree of Life. We share most of our DNA, and our history, with other species; and therefore we cannot be separated from nature, nor can nature and its species be separated from us. If our morals and ethics dictate that every human has an equal right to live, I cannot see the logic of why we should not ascribe similar rights to non-human species.

The idea that nature and species have their 'own rights' has long roots in many indigenous cultures, and has been gaining increasing traction in recent years. Under the umbrella of 'Rights of Nature', organisations around the world are trying to give mountains, rivers, oceans and their biodiversity legal rights to 'exist, persist and regenerate'. This contrasts with the status quo that nature is the 'property' of whoever owns the land, which effectively gives them the right to destroy it if they wish (unless the land is under some form of legal protection). Ecuador was the first country to recognise the 'Rights of Nature' in its constitution in 2008. Since then, similar successes have been achieved across the world, including in Bolivia, New Zealand, Mexico, Uganda, Bangladesh and India. In some cases, such rights have been adopted in local rather than national law, including in dozens of cities and counties in the USA.

These are still isolated victories, though, unknown to most people, and with limited power when things get really serious. It is also clear that many people, including some of our most powerful political leaders, don't share the view that biodiversity has an intrinsic value that is worth protecting. In 2019, the Amazon experienced some of the worst fires in decades, set alight by farmers who wanted to expand their land for growing beef or soy. It was obvious that what was happening could be strongly linked to the controversial plans of the outspoken Brazilian president, Jair Bolsonaro. His plans to 'develop the Amazon' involved promising no punishments to land grabbers, claiming that indigenous communities already had 'too much land', and encouraging the purchase of weapons by farmers. In response, I wrote a strongly worded blog for Kew, entitled 'The Amazon is burning. Will the world just watch?' where I articulated the consequences of destroying the largest, oldest and most diverse rainforest on Earth, and this got me invited to several radio interviews and debates around the world. The president responded to escalating criticism shortly after at a Climate Summit held at the United Nations, where he clarified that the Amazon is Brazil's property – not humanity's nor the world's – and if the country wanted to use it for 'economic development' it was entitled to.

Being a Brazilian national, with the Amazon a primary focus of my research and admiration, I reflected on how laws can be designed to suit the lawmakers, often fussing over trifles while ignoring much more serious matters. When I was preparing my first trip to the Amazon for my Master's project, I needed to obtain permits for collecting plants to make herbarium specimens. At that time, I knew that many researchers didn't even bother to apply for a permit, since they knew it was so bureaucratic they would never get one. But I wanted to do the right thing (and avoid ending up in jail), so I started the application six months before the trip. I responded to increasingly pedantic requests for information, such as what species I wanted to collect from which exact locations, which, if I had known, would have made my trip pointless. Nevertheless, after filling in piles of paperwork from multiple government agencies, I finally got a reply a week before my fieldwork was due to start: 'The President of Brazil needs to approve your request.'

In effect, a young Brazilian student was being denied permission to collect a handful of branches from poorly known species for his research, with no environmental impact whatsoever. At the same time, wealthy farmers and landowners were entitled to slash and burn vast areas of rainforest each year, with full legislative support. My sense of powerlessness and frustration couldn't have been any greater. At the very last minute – half an hour before leaving for the airport on the way to the Amazon – I got a phone call from Brazil's Environmental Agency, saying that they had approved my request and would send the paperwork by fax directly to the hotel at my destination. I was relieved, but still perplexed at the absurdity of the whole situation.

Despite resistance from some political leaders and a lack of recognition in our legal systems, I hope the arguments I've presented in the last three chapters show that the imperative to value and protect the world's biodiversity is undeniable, from both selfish and altruistic perspectives. In the face of the massive environmental threats that our generation is responsible for creating, we must focus all our attention on saving our planet's species and ecosystems, before it is too late. To achieve this, we must first identify and understand the primary drivers of biodiversity loss.

PART THREE

THE THREATS TO BIODIVERSITY

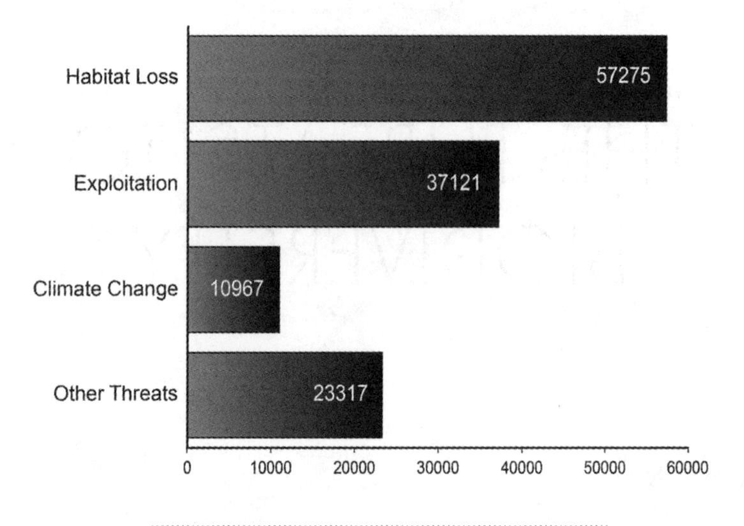

Figure 14. Major threats to biodiversity. This section explains how the loss of natural habitats, direct exploitation of species and climate change are posing the biggest threats to the world's species. Other hazards and risks – including invasive species, pollution and diseases – add considerable pressure and act synergistically. These numbers derive from conservation status assessments carried out by the International Union for Conservation of Nature and its partners. The threats to most species haven't been assessed yet.

Silently, a black hole is engulfing our biosphere – the thin layer of life that supports all life on Earth and sets this planet apart from all other known objects in our vast, ungraspable universe. But rather than an irreversible celestial phenomenon, this black hole is nothing less than our infinite greed. Human consumption, particularly in wealthy societies that have long profited from systemic global inequalities, is peeling off the biosphere one chunk at a time. Just as we live our normal lives and maintain a level of resource utilisation that is anything but sustainable, biodiversity is disappearing at rates never previously recorded in human history. A species of ape that calls itself 'wise' – Homo sapiens *– has yet to do justice to its name. Rhino after rhino, orchid after orchid, a million species are now predicted to be at risk of extinction. How did we get to this ghastly situation, which is completely undermining our own chances of survival as well as those of so many species around us? The causes are manifold (Fig. 14) – as we'll now find out.*

Figure 15. A scarlet macaw. This is one of the numerous amazing birds that inhabit the Pantanal wetlands of South America.

CHAPTER 9

HABITAT LOSS

One of the first family trips I remember was in the 1980s to the Pantanal – the world's largest wetland system, just south of the Amazon. From home, it took us two days to get there. Gradually, the landscape changed from São Paulo's urban environment to a mosaic of natural savannahs interspersed with rivers, ponds and forest patches. My brother and I had a competition over who could spot the most jabirus: a beautiful, huge, black-headed stork with a red neck collar and white body. We laughed when we shouted its Portuguese name every time: 'Tuiuiú!' I lost track after a few hundred, and then I got distracted after spotting so much other wildlife: sunbathing caimans, herds of capybaras (the world's largest rodent), flocks of scarlet macaws (Fig. 15), toucans, birds of prey, and much more. I later learned that the Pantanal is known globally as a unique haven for biodiversity. I marvelled.

Some 15 years later, when I took my wife-to-be to Brazil for the first time, I wanted to show her the best of my country. The Pantanal would naturally be our first stop. This time, however, the landscape wasn't as I remembered it. It took us much longer to start seeing wildlife from the road, as cities had swollen in size, and large monocultures of soybean had expanded far into the savannah. And as I write these lines, now 20 years since my last visit, I know for certain that the Pantanal isn't what it used to be. In 2020, in a single year alone, around a quarter of its entire area was set on fire by farmers who wanted to expand their land for growing soy or raising cattle.

Unfortunately, the Pantanal is no exception: in most parts of South America, and indeed everywhere else in the world, our natural ecosystems – forests, wetlands, savannahs, grasslands, seabeds, coral reefs – have undergone massive changes and degradation. Both on land and at sea, habitat loss has become the world's primary driver of biodiversity loss.

Humans have modified our planet for millennia. Archaeological and paleoecological evidence – such as human artefacts, pollen and charcoal – are increasingly challenging the idea of 'untouched' and 'pristine' habitats, showing substantial alterations of the vast majority of ecosystems by human actions since at least 12,000 years ago. However, never before have our activities been so intense, and so devastating, as now. The rapid biodiversity loss that we are seeing today is primarily explained by the intensification of how we exploit nature, which is substantially different from the generally much more sustainable way that indigenous, traditional and local communities have tightly interacted with nature throughout history. In most regions, the big changes started very recently – in conjunction with what is called the **Great Acceleration.**

The Great Acceleration is a period of rapid and drastic changes that began around the 1950s. Since then, almost all measures of human activity – population growth, greenhouse gas emissions, food production, pollution, water usage and many others – have drastically increased*. Many of the modern land changes have been for cultivation, livestock raising and plantations, to meet the increased requirements of the world's human population – which has been both expanding and demanding more per person, with vast differences across societies. In South America, over 70 per cent of deforestation in recent decades has been for cattle ranching, and a further 14 per cent for growing animal feed and other commercial cropping. Soybeans grow fast and are rich in proteins, making them the ideal fodder for the low-cost production of beef, pork and poultry around the world. But agriculture uses more freshwater than any other human activity – with nearly a third of its use required for livestock alone – and crop monocultures require vast amounts of pesticides, contaminating the environment far beyond their borders. In the vast fields where soy is planted, almost no other species survive.

* Recent data seem to indicate a slowdown for several of these trends over the last years, which might have been further exacerbated by the COVID-19 pandemic. However, climate change continues to accelerate, and certain regions such as Africa continue to show rapid human population growth.

In Southeast Asia and increasingly in tropical Africa, the key drivers of deforestation are oil palm plantations. Like soy, oil palms grow quickly and are cheap, and their demand has increased drastically – now found in nearly any product you see in the supermarket (often disguised under the cryptic name 'vegetable oil/fat'): margarine, chocolate, cookies, ice cream, noodles, shampoos, detergents, lipstick, the list goes on – they can all contain palm oil.

In the ocean, similar drastic changes have taken place to seabeds, whose fauna has been devastated in certain places by the intensification of trawling, underwater mining and other forms of physical and chemical damage. This is a form of 'silent violence', unnoticed by the vast majority of people and authorities and much more difficult to survey and rescue than deforestation or other changes on land, which can now be monitored through satellites and other remote-sensing technologies in almost real time.

It's easy to understand why biodiversity is lost when its habitat disappears, especially those species that are restricted to small areas or very particular habitats, like lemurs in Madagascar (which have evolved into different species, sometimes confined to single valleys) or pandas in China (which used to be much more widespread in the past, before being decimated by human activities). Among vertebrates, the most striking known example is the Devils Hole pupfish. With an average length under 3 centimetres (about 1 inch), this blue-coloured species has its entire **range** in a limestone pool that is 22 metres (72 feet) long by 3.5

metres (11 feet) wide. In the 1960s and 1970s, local farmers began to extract underground water to irrigate their crops, which caused the water level in the hole to drop, reducing the pupfish's habitat even more. In 2006, there were only 38 pupfish remaining in the wild. Although the population size has increased since then, it remains a critically endangered species. As the surrounding landscape in southern Nevada, USA, is part of the Death Valley National Park and inhospitably hot (when I was there in 2018, temperatures reached 50°C, over 120°F), there isn't really anywhere else the species could naturally go.

Sometimes, the effect of habitat loss can also be felt from afar. When I first came to live in Europe in the late 1990s, I remember what invariably happened when we drove a car in the countryside, especially on dirt roads: insects would crash against the windshield, in such profusion that we often needed to turn on the wipers. I'm sure many others will remember this, but in just a matter of years this phenomenon was almost gone, and my first child, born in 2004, has never experienced this. In 2017, researchers documented a reduction of over three-quarters of insect biomass (their total weight measured from traps) over the course of just 27 years, and a reduction of over a third of species in just 10 years. Unexpectedly, all those losses were measured in existing protected areas, which were meant to shield the local biodiversity. Although the exact causes of these losses remain debated, many believe they are explained by the deterioration of habitats across large landscapes, over which insects and other organisms could formerly move freely. This drastic decrease in insects has had a negative knock-on effect for the animals that feed on them, such as birds, bats and dragonflies.

Another aspect related to habitat loss derives from the species-area relationship mentioned in Chapter 1 (Fig. 4). As we saw, the larger an area is, the more species it will accumulate over time, as a result of more opportunities for the rise of new species, or speciation, and **colonisation** of these habitats by species living elsewhere. Unfortunately, the opposite is also true: if you decrease the area of a particular habitat, it will inevitably host fewer species over time. So, many of the forest fragments left today, which represent remnants of much larger ecosystems, probably contain a high but unknown degree of **extinction debt**, which means there are more species in the area than it can sustain in the long term. This is an area of active research and many questions remain, but the insights gathered so far are pretty scary. We don't quite know how long it may take for species in those fragments to die out, often as a result of low genetic diversity, insufficient amounts of food, or higher risk of diseases.

Habitat loss is not only affecting species diversity within well-known ecosystems like tropical rainforests, but also ecosystem diversity itself. There is a disproportionate impact on ecosystems long regarded by many as either worthless or 'blocking the way' for crops and other uses, such as wetlands and grasslands. Asia, for example, has lost two-thirds of its natural wetlands since 1945, and 84 per cent since 1900.

The research of Cédrique Solofondranohatra, a skilled Malagasy scientist working with Kew and other collaborators, has shown that many grasslands long regarded as 'manmade' in Madagascar are in fact natural and ancient, mirroring previous misconceptions in mainland Africa and Australia. Clues come from the high diversity of grass species unique to the island, and the fact that **morphological** features and species communities of those grasses show clear indications of evolutionary adaptations to fire and against grazing animals; none of these could have evolved under the relatively short period that humans have been present. Although many native grazers are now gone – such as Malagasy hippos, elephant birds and large lemurs, all hunted down to extinction by humans – the introduced ox (a variety called zebu) helped to keep those areas open. This is important, because early colonists and even conservationists have assumed that rainforests should be the primary vegetation type of most of the island's interior, rather than just confined to the eastern and northern coastal areas as today. Their assumptions were based on biased views of nature from mainland European forests, which lack natural fire regimes.

The lack of appreciation for grassy ecosystems, which in Madagascar has persisted for a long time due to the country's long isolation from international science, combined with our limited understanding of their natural extent and diversity, have contributed to their existence now being severely under threat, alongside the iconic rainforests and in fact every other ecosystem on the island.

While people in wealthy countries often see Madagascar's unique biodiversity as something that needs to be 'saved', for millions of Malagasy people, the degradation of ecosystems and the loss of local biodiversity primarily mean a steady decay in their ability to derive from nature the most fundamental needs for subsistence: wood for cooking, heating and shelter, clean water, food and medicines. If conservation work in the country is to be successful, it therefore needs to address the root causes of biodiversity loss, including poverty and food insecurity. Beyond the immediate impacts on biodiversity, habitat loss also influences local and regional climate, and by releasing vast amounts of carbon dioxide into the atmosphere it contributes to global warming that affects all of us. The protection of Madagascar's remaining biodiversity, alongside the restoration of degraded ecosystems, hold the potential to deliver positive outcomes for both people and nature.

It is also in Madagascar that we see some of the most striking effects of the second main driver of biodiversity loss: the exploitation of individual species.

CHAPTER 10

EXPLOITATION

If the degradation of biodiverse ecosystems around the world wasn't enough – such as slashing and burning the Amazon to raise cattle and grow soybeans – the killing and exploitation of both animal and plant species is putting many at risk of extinction. The increasing demand on meat consumption from wild animals, for instance, has a direct effect on extremely rare primates in Africa, as is the felling of rosewood in Madagascar for furniture purchased in wealthy countries putting those trees on the brink of extinction. In the ocean, up to three trillion fish are caught every year, making the unsustainable exploitation of species the biggest driver of biodiversity loss in the marine realm.

Sometimes, hunting wildlife is the immediate and perhaps only solution for hungry families to survive under extreme poverty. During my research around the world, I've met with many hunters who lacked any source of income, so they didn't really have an option. Increasingly, though, hunters are tourists who kill for pleasure, making it totally unnecessary and indefensible. The organs of many threatened species are mistakenly believed to have aphrodisiac or curative properties, like rhino horns and double coconuts from the Seychelles. The capture of wild animals for the pet trade is another example and has become a major and growing threat. Alice Hughes, a prolific scientist at the Xishuangbanna Tropical Botanical Garden in China, estimated with her colleagues that nearly 4,000 species of reptiles – over a third of all known species – are now in trade. Of these, 90 per cent are captured from the wild, and three-quarters are not covered by international regulation, such as many threatened and range-restricted species, especially in Asia.

A big problem with many exploited species is that they are rare, either naturally or due to human actions. Orchids are a good example where casual picking by local people, and targeted harvest by commercial traders, have had a substantially negative impact on the plants' numbers and their survival. During my PhD training, I went on a botany excursion to the Mediterranean island of Crete. In one of our hikes over the limestone mountains, I stumbled on a 20-cm-tall (8 inches) plant with white flowers just a few metres away from the main trail. I knew it was an orchid but had no idea which species it belonged to. When I showed it to Arne Strid, our teacher and expert on the Greek flora, he immediately knew it was *Cephalanthera cucullata* (Fig. 16): one of the world's rarest plants. He had seen it in exactly the same spot many years before, and very few individuals of that species have been recorded since then, despite much searching by botanists. Had I

Figure 16. The rare orchid *Cephalanthera cucullata* from the Greek island of Crete. This is one among nearly 10,000 plant species documented as threatened due to direct human exploitation (often in combination with other factors).

dug it up as a souvenir, I might have exterminated the last individual of an entire population and its genetic diversity, eventually contributing to the extinction of the species.

One common trick by sellers of threatened species – which are often highly priced in the black market – is to use another species' name in any required paperwork. I was once collecting research permits at a government office in the Dominican Republic, before heading off to fieldwork. While waiting for my turn, I saw two men being escorted by a police officer. They placed a large paper box on the ground just behind me. I looked inside and saw over a dozen young parrots. The colour patterns were unmistakable: they all belonged to the same endangered species endemic to the island of Hispaniola. Yet the officer told me the men had claimed it to be a common and widespread species whose trade wasn't restricted. He asked my opinion about the species; although I'm not a bird expert, I always carry a bird guide in my backpack, so it was easy to confirm the officer's suspicion.

Whether deliberate or not, erroneous species identifications like the Dominican parrots are a huge challenge for regulatory authorities, who often lack the expertise or tools to verify the identity and provenance of species in trade. Perhaps the biggest challenge of all is timber: the wood used for furniture, building material, floors, musical instruments, fuel, paper and several other crucial products. The timber sector employs over 50 million people and generates some US $600 billion annually. The demand is expected to quadruple by the middle of this century. This trade includes highly fashionable wood such as rosewoods and mahogany. How can we tell apart a threatened from a non-threatened species? How do we distinguish wood of a threatened species that is sustainably cultivated, from wood harvested in the wild?

To tackle these critical and difficult questions, my colleagues Peter Gasson and Victor Deklerck at Kew are working together with a network of partners* to build one of the world's largest collections of wood samples, both in terms of species and geographic origins (Fig. 17). They're now using an array of techniques to identify unknown samples. To recognise the species, they take thin slices of wood and examine them under the microscope. By studying the overall appearance, number and arrangement of the different types of wood cells, and comparing the structure of the slices with reference samples of known species stored in our collections, they can match the unknown sample. The use of image-recognition algorithms – a powerful application of artificial intelligence – is now speeding up and improving this process. Other possibilities being pursued are to compare the chemical profiles of unknown wood samples with reference profiles, and to sequence DNA from the unknown samples, and then compare it with previously published reference sequences. For example, another colleague – William Baker – has been using DNA to help IKEA identify the species of rattan cane used in several of their furniture pieces, as a first step to ensure

* For more information, visit https://worldforestid.org.

they come from sustainable sources. Rattan is a collective name for hundreds of species of climbing palms, particularly diverse in Southeast Asia; in contrast to most timber species, they are still primarily harvested from the wild rather than cultivated, so it is crucial that no threatened species are used.

Once we have identified the species of wood, we usually want to know where it comes from. For this, we examine the chemical signature of samples. Luckily, the distribution between different isotopes (variants of individual chemical elements) in the wood depends on where the tree grew, due to factors such as rainfall, temperature, topography and local geology. Although it's not always possible to locate the exact provenance or identity of samples, at least until we have a more complete set of reference samples, in most cases the results are enough to verify claims made in the seller's documentation. This technique is increasingly allowing authorities to seize illegal shipments at borders. By cutting the supply chains, our hope is to diminish the demand for illegal exploitation in the first place. Worryingly, the work done so far has shown that some 40 per cent (!) of all timber traded internationally is probably illegal. This is a key reason why roughly a third of the world's estimated 73,000 tree species are now threatened with extinction.

Besides the obvious loss of species diversity, direct exploitation also leads to important losses in both genetic and functional diversity. This is because we often target the most valuable individuals to us, rather than 'random'

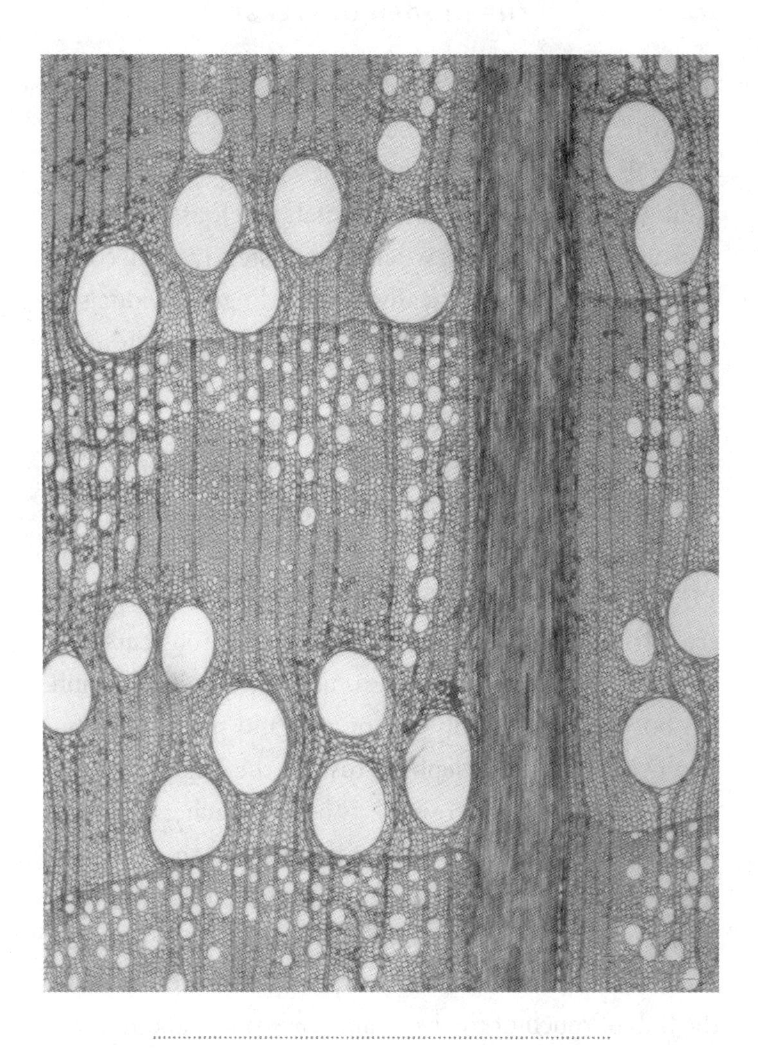

Figure 17. Cross section of a wood sample (here the common oak). The large diversity of shapes, disposition and sizes of vessels and other structures in wood, combined with modern chemical analyses, are allowing scientists and authorities to identify the species and geographic origin of wooden artefacts, such as furniture and musical instruments. A substantial proportion of the international timber trade appears to be illegal.

individuals in nature. Trophy hunting is a classic example where humans may be counteracting evolution by 'natural' selection, as originally described by Darwin. By killing the most impressive animals, such as those with unusually large antlers, we have consistently reduced the chances of those genetically outstanding individuals to reproduce, leaving behind mostly average animals and a reduced gene pool.

The differential level of hunting across different regions and time periods, often linked to colonial history and global trade, has led to some species losing entire populations and the inherent diversity they contained. In 1533, a Portuguese ship went missing off the coast of southern Africa. The ship was recovered nearly 500 years later during a mining project, in astonishingly good condition. On board were not only lots of gold and silver coins, but also over a hundred elephant tusks – the largest cargo of African ivory ever discovered. When researchers analysed the DNA and stable isotopes of those tusks, they saw that they belonged to forest elephants from West Africa, many of them belonging to populations that no longer existed. Their disappearance was a consequence of the region being the hub of much of the ivory and slave trade, and in more recent times the competition with a rapidly expanding human population, habitat encroaching due to the spread of agriculture, exploitation of natural resources and civil unrest. Since elephants play a huge role in ecosystems (by opening up forests and allowing more light to reach the ground, as I mentioned in Chapter 7), the loss of those populations meant the loss of those vital functions in that region, as well as considerably reduced overall genetic diversity among African elephants.

The loss of functional diversity as a consequence of human activities is probably most evident on isolated islands. Every time humans arrived on a new island, they often encountered many 'naïve' species of animals that had never seen a predator before. When Dutch colonisers first settled on the island of Mauritius in 1638, the island abounded with giant tortoises. These were, however, quickly slaughtered in high numbers by the Dutch, and were used as a food source for the people and their pigs, and were used as a source of fat and oil. By 1700, the species was probably already extinct on the island. Similarly, since most islands naturally lacked large predators, and flying is an energy-demanding activity, many species of island birds lost their ability to fly over time, making them an easy catch. In a study led by Spanish researcher Ferran Sayol, we estimated that at least 581 bird species have gone extinct over the last few thousand years, a large proportion of which were restricted to islands. While our ancestors who colonised those islands tended to ignore small and fast-moving species like passerines, they hunted down all the big, meaty ones to extinction, including several large species of moas from New Zealand and the elephant bird from Madagascar. The heaviest dove that ever existed – the iconic dodo from Mauritius – was said to have an unpleasant taste to humans, but they did not survive the cats, rats and pigs that people brought with them.

In some cases, such as with the Arctic fox in northern Scandinavia or sharks in coral reefs, humans haven't managed to kill all individuals of a species, but have left so few behind that they are particularly vulnerable to another major threat: the world's increasingly warm and unstable climate.

CHAPTER 11

CLIMATE CHANGE

Climate change is one of the biggest challenges that our societies face. It affects global food production, water availability and health, and leads to sea level rises that over time may force the displacement of hundreds of millions of people living in coastal regions. But when it comes to biodiversity, the reality is that it is only the third biggest threat, after habitat loss and the direct exploitation of species. That isn't to say that climate change isn't important – it is. Now that so many natural ecosystems have already been destroyed, and many species have already had their populations decimated, climate change is likely to become an increasing threat to our remaining species over the next decades.

Many people equate climate change to the steady increase in temperature over time around the world. This is indeed one of the most noticeable effects – think back about how the summers and winters of your childhood were probably quite different to the ones today. However, there is more to it than that. Climate change also refers to the marked differences we are seeing in rainfall around the world, with some regions like Australia and the Mediterranean becoming increasingly dry, whereas others, especially around the Equator, are becoming wetter. And if there have been any doubts about the underlying causes of those changes, they're now gone: undeniably, the cause is us – through emissions of greenhouse gases such as carbon dioxide, methane and nitrous oxide. These gases are produced from a variety of human activities, in particular the production of electricity and heat, agriculture, transportation, forestry and manufacturing.

Most species have a narrow climatic tolerance and thrive best within a small temperature range. We're no exception: in offices, research shows that the ideal ambient temperature is almost exactly 22°C (72°F). Increase it by a few degrees, and our ability to make complex decisions drops; decrease it by a few degrees, and productivity decreases. Although these calculations are based on a stereotypic 70 kg (150 lb) male (women usually prefer a few degrees higher), they apply regardless of where people come from. So, perhaps not surprisingly, for thousands of years, people have been living within a small subset of the planet's full climatic range. Although we have occupied regions ranging mostly between 8–28°C (46–82°F) in mean annual temperature, there has been a preference for sites around 13°C (55°F), characteristic of today's Beijing, Milan, Wellington or New York City. Although it's difficult to establish causal links – since other factors are likely to be involved, such as the potential for agriculture and avoidance of tropical diseases – it's nevertheless remarkable how closely associated we are with climate.

As the climate changes, species that cannot survive the new conditions have two options – either adapt to the new conditions, or move to a new place with a more hospitable environment. If they don't, they will become extinct. Some species and populations within species are indeed showing signs of rapid adaptation and benefitting from a warmer climate, like some lizards studied by my Brazilian colleague Fernanda Werneck at the National Institute of Amazonian Research (INPA) in Manaus. But unfortunately, most species, including humans, are notoriously bad at adapting biologically. A lot of my research has gone into understanding just how species responded to past increases in global temperature, but in all those cases the changes took much longer than we're seeing today. In fact, it's been estimated that species might need to adapt up to 10,000 times quicker than they've ever done in the past, which isn't going to happen, at least for many species (just how many is a topic of scientific debate).

Moving elsewhere is therefore the main hope for species threatened by climate change. In the past, moving from A to B wasn't as big a deal as it is now – we've fragmented most ecosystems on the planet, adding new barriers to free movement, including cities, roads and croplands. In some countries, building forest bridges over big roads is helping to facilitate safe passageways for animal movements, and conservationists are manually helping a few iconic species (mainly mammals and amphibians) to reach new areas. However, most species don't get that luxury.

A beam of hope is offered by the world's mountains. Long ignored by humans due to the challenges of growing crops or harvesting timber in rugged landscapes, mountains around the world have been conserved disproportionately to flat, easily cultivated areas. This is good news for two reasons. First, mountains are naturally very biodiverse: although they occupy a mere eighth of the world's land area, they are home to about a third of all terrestrial species, thanks to their complex concentration of many different habitats in one place. Second, species living on mountains need to move much shorter distances – a short jump uphill – to track their optimal temperature, whereas species living in flatter regions may need to move hundreds of kilometres (usually away from the equator) to find a climate similar to what they are used to. In the Andes, dozens of plant species have been able to move uphill by hundreds of metres since Humboldt climbed its mountains over 210 years ago, tracking their optimal climate and vegetation zone.

The bad news, though, is that not all species are able to move uphill as fast as temperatures are increasing, and even if they do, there is always a final limit: the summit. Although scientific evidence is limited, often due to the lack of reliable historical records, it seems that many mountain species might be lagging behind in speed. To make things even more complex, many species that show strong ecological interactions with others – such as plants and their specific pollinators – need to move together. So even though mountains provide future refuges for species currently at lower elevations, it is crucial to conserve corridors along elevation gradients – from low to high altitudes – to increase the possibilities of long-term survival and free movement of species. This can work in many regions, but not all – like Australia, which is largely flat with the highest mountain reaching 2,228 metres above sea level. While biological corridors connecting natural habitats in lowlands are equally important to improving the long-term survival of species, by allowing them to exchange genes and maintain appropriate population sizes, they may not offer the same buffer against climate change as those in mountains.

Polar species already living at climatic extremes are among the most vulnerable, since they often have nowhere else to go as their habitats melt away. Polar bears, arctic foxes, walruses, narwhals and many other animals on land and in the sea have an intimate relationship with ice and snow. During fieldwork in the mountains of northern Scandinavia in the early 2000s, I helped my colleague Ulf Molau and others to capture lemmings – small and exceedingly cute rodents that can occur in huge numbers in certain years – to assess how climate change might be affecting their weight and number of babies. Even then, we were beginning to see significant changes in their survival. The increasingly warm winters brought more rain and melted down snow, which soon froze and created a hard layer of ice that the small animals couldn't break through, starving to death. Even the much bigger reindeer faced a similar problem there, not being able to reach down to the lichens growing beneath the snow, which constitute the majority of their diet during the winter.

Some ecosystems are particularly sensitive to climate change, with coral reefs (Fig. 18) one of the most extreme and alarming examples. Their survival at present is on a knife-edge, due to their sensitivity to heat. This is because as water temperatures rise, corals expel the symbiotic (co-living) algae within them. This causes the corals to become

Figure 18. A healthy coral reef. These fascinating and biodiverse ecosystems consist of strongly inter-connected species. Having existed for tens of millions of years, climate change caused by human activities is now posing severe threats to their survival, with a nearly full collapse predicted unless drastic measures are put in place.

completely white by exposing their bare carbonate surface – they are said to 'bleach'. Over time, if stress conditions continue, both the corals and the algae may die, since the algae are essential to the survival of corals.

It's tempting to think that a mere 0.5°C won't make any significant difference, but this couldn't be further from the truth. If global warming stays within a maximum of 1.5°C (2.7°F) increase over this century, as aimed for under the Paris Agreement, only some 10 to 30 per cent of all the world's shallow-water coral reefs will be expected to survive. This is already an extremely bad prospect, but it is unfortunately a best-case scenario. If we instead reach a 2°C (3.6°F) increase – the minimum that most projections, based on current trends, are pointing towards – less than 1 per cent of coral reefs are expected to survive. This is an incredibly dire future to face, given the extreme diversity in those ecosystems which evolved over tens of millions of years and today provide manifold benefits to over 500 million people around the world – as sources of fishing, tourism, coastal protection, medicines and more.

However, there's an additional threat linked to carbon dioxide emissions: **ocean acidification**. At least a quarter of the 40 billion tons of carbon dioxide released by human activities each year is absorbed by the sea. This, alongside the absorption of heat from the atmosphere, are amazing and underappreciated services that oceans provide to us, and which help mitigate the immense damage we are inflicting on our planet. However, this comes at a high price, because all that carbon leads to a marked increase in the sea's acidity. Since 1850, average acidity in the world's seas has increased by some 30 per cent, a level that could triple by the end of this century. As shown by my French colleague Sam Dupont and others, there is now a vast diversity of species, including those with skeletons or shells containing calcium carbonate, that are sensitive to acid, such as starfish, brittlestars, mussels, oysters and sea urchins. Keeping their bodies' acidity to the right levels is critical for marine organisms, and ocean acidification means they have to expend extra energy to maintain these levels, limiting their growth and weakening them, and sometimes causing them to die. As well as reducing the number of incredible species in the sea, this impact can have important knock-on effects across the ocean's complex food web.

Sometimes the impact of climate change on species is rapid and noticeable. A typical example is how global warming is changing the **phenology** of species, which is the timing of seasonal events, such as when plants flower, set fruits and shed their leaves; when certain birds migrate; when frogs and toads lay their eggs in water; and when fish spawn, among many other phenomena in nature's recurrent cycles. In a few cases, phenology is independent of climate: this is the case for oats, rice and soybeans, whose flowering time is controlled by light receptors that respond to the balance between day and night. But for most species, phenology is largely regulated by climate. Historical records are invaluable, to document whether certain phenological observations are 'normal'. This is what motivated Japanese researcher Yasuyuki Aono to search through documents written by emperors, governors and monks, who kept detailed records of when the cherry trees or 'sakura' would blossom in the city of Kyoto – a major annual cultural event. His data, which tracked flowering times back to the year 812, showed that 2021 witnessed the earliest ever flowering peak, on 26 March. Plants that start flowering earlier than usual may miss their pollinators, such as insects, which may not have left their larval stages or pupas yet. When those insects eventually emerge, their preferred flowers will have already withered, and so they may not get enough food and thus die. Other such asynchronies may lead to plants not having their seeds dispersed, or animals laying eggs too early in the season, becoming more susceptible to freezing temperatures or periods of drought.

A steadily warming climate is one challenge, but extreme weather events are quite another. Overnight, they can destroy an entire ecosystem. Around the world, we're seeing that heat waves, droughts, fires, flooding and hurricanes are increasing in both frequency and intensity. Australia is a striking example of where this is happening in front of our eyes. In 2016 and 2017, heat waves led to the bleaching of the Great Barrier Reef, killing around half of its corals. In 2019, more than 10.6 million hectares of land were affected by bushfires across the country, which Australian scientist Chris Dickman and colleagues estimated to have killed 2.5 billion reptiles and 143 million mammals. Although much of Australia's terrestrial **biota**, such as eucalypts, evolved under a regime of regular fires, the extreme temperatures in 2019 both amplified the fires and allowed them to reach areas not previously burned. At Kew, we store collections of seeds for nearly 9,000 Australian plant species, so we were able to help our partners in providing some material for restoration – but this is a long and difficult job, and a drop in the ocean that cannot fully replace the biodiversity lost during that tragic event.

Biodiversity loss and climate change are intertwined global challenges. When ecosystems are degraded, carbon is released and rainfall patterns are disrupted; and as climate changes, so does the diversity and distribution of species, and the health of ecosystems. Breaking this vicious loop will require tackling both crises simultaneously. We have to do everything we can to halt and mitigate climate change, and also find ways for us humans and the natural world to adapt more rapidly to the irreversible changes that we have already set in motion. As if habitat loss, species exploitation and climate change weren't bad enough for biodiversity, their impacts are accompanied – and sometimes even amplified – by several other threats, which I'll now set out.

CHAPTER 12

OTHER HAZARDS AND DANGERS

INVASIVE SPECIES

There are several other drivers of biodiversity loss that can have a significant impact on our hidden universe. One of them is the threat from **invasive species**, like the introduced North American beavers I helped local conservationists to hunt down some years ago on the world's southernmost inhabited island – Isla Navarino in Chile, a place of incredible natural beauty. Beavers were introduced in 1946 to kick-start a fur trade, but they spread rapidly amongst the islands around Tierra del Fuego. As **ecosystem engineers**, or species that modify their environment significantly, the beavers transformed large areas of closed southern beech forests to grass- and rush-dominated meadows, building dams and dens, and altering stream and nutrient cycles (Fig. 19).

The slow-moving water behind their dams became attractive to two more introduced invasive species – mink and muskrats, with minks in particular feeding on native geese, small rodents and other native wildlife. This has led to what researchers have described as an 'invasive meltdown process', where one invasive species amplifies the impact of others, leading to a massive environmental effect: degradation of the species-rich, natural habitats on the island, the decrease of population sizes of several native species, and disruptive changes in nutrient cycling, water streams and soils. This is why my fellow conservationists had to resort to such drastic measures as killing beavers – in themselves lovely creatures – in an attempt to restore the local biodiversity and ecosystems.

The impact of invasive species can be very fast. My wife and I met in a diving school in Honduras, and ever since we moved to Sweden in 1999, shortly after my father had passed away, one of our favourite summer hobbies has been to snorkel on the country's acclaimed west coast. I quickly familiarised myself with the local species and enjoyed exploring the diversity of shallow-water ecosystems. But in the summer of 2007, everything changed. Suddenly, what used to be diverse communities of algae, crustaceans and molluscs on the rocky surfaces of the region's many islets became dense mats of a large oyster I had never seen before. I took photos and soon found out that these were Japanese oysters – a species native to the Pacific coast of Asia, which has been introduced to many parts of the world as a food commodity. The problem is that it is a particularly aggressive species, outcompeting the native fauna, such as the blue mussel, and leaving little

Figure 19. Beavers at work in the world's southernmost inhabited island, Isla Navarino. The introduction of this exotic species from North America is leading to a major deterioration of the fragile natural ecosystem.

room for others. How exactly the species came to Sweden remains a mystery. It may have arrived through water currents from other regions where it is grown in Europe, or – as is often the case with marine invasions – it may have arrived in the ballast water of ships. Interestingly, Swedes had once tried to grow it in the same region in the 1970s, but the colder water temperatures at the time meant that the species could not reproduce. This shows how climate change is facilitating the spread of many more invasive species around the world, and exemplifies how different drivers of biodiversity loss can interact.

MULTIPLE FORMS OF POLLUTION

It is also in the sea that we see some of the biggest effects of human activities on land, and yet another major hazard to species: pollution, consisting of both rubbish and chemicals. Among the many types of litter, few can match the long-lasting and damaging effect of plastics – the most common form of debris in the oceans. The first synthetic plastic was created in 1907, and it was not until the end of the Second World War that mass production really began. On average, an astonishing 50 kg of plastic is produced per person each year – 99 per cent of which is produced from oil and gas.

Packaging materials account for about half of the plastics produced, and are among the most commonly identified types of litter. Plastic bags, bottles, caps and food wrappers end up in streams and rivers, reaching seas around the world where sea turtles, whales and other animals mistake them for food. It is estimated that 99 per cent of all seabirds will have ingested plastic waste by 2050, many of them eventually dying from it. Less than 10 per cent of all plastic is currently recycled, meaning that most of the plastic ever produced has not been reused; some remains in use in our **technosphere**, which is the human-built part of the world and comprises all of the technological objects manufactured by humans, such as machines, roads, railways and buildings. Much of the rest remains in landfills or the wider environment, where it takes hundreds of years to decompose.

In the environment, plastic then breaks down into increasingly smaller bits (**microplastics** and nanoplastics), which in the water are taken up by many different species of plankton, gradually moving up through the food chain to shrimps, fishes, birds, and then mammals like seals, bears and humans. In the body, they damage cells and can induce inflammatory and immune reactions, but the reality is that we still have almost no idea about what other negative, potentially substantial effects they could have on us and other species.

Chemical pollution can be equally damaging to wildlife. These include the over 350,000 manmade chemicals and chemical mixtures produced so far, with some 40 new sorts of chemicals being synthesised *each hour*, many of which will become available on the market and make their way into the environment. The production and use of chemicals vary hugely, and their presence is ubiquitous:

from poisonous mercury leaked out during gold mining in the Amazon, to sulphur dioxide released into the air when coal is burned, causing acid rain, to nitrogen oxides produced as by-products of motor vehicles such as cars and aeroplanes. In aquatic environments, nutrient run-off from fertilisers used in agriculture, containing elements such as nitrogen and phosphorus, promotes the bloom of cyanobacteria that are toxic to many animals, suppresses growth of many aquatic plants by blocking sunlight and creates oxygen-free dead zones in the bottom of seas and lakes. At the same time, synthetic hormones from birth-control pills are flushed down the toilet and end up in lakes and seas, disrupting the reproduction of fish species, feminising males and affecting egg development in females. It is perhaps not surprising that pollution, in combination with direct exploitation and habitat loss, has led to a global crisis among freshwater fishes – one in three species is now facing extinction, and the population sizes of large species studied in detail have shrunk by an average of 94 per cent over the last 50 years alone.

On land, chemicals have had at least as much of a negative impact on biodiversity as they have in the sea. The discovery that the synthetic molecule DDT (dichloro-diphenyl-trichloroethane) was highly effective against insects – a finding that was awarded a Nobel prize to Swiss chemist Paul Müller in 1948 – led to its widespread use in agriculture for decades after the Second World War. However, it eventually became clear that DDT (and the structurally very similar molecules DDE and DDD) accumulated in animal tissues, particularly among birds of prey, waterfowl and songbirds. The chemicals caused their eggshells to become thinner and break, leading to a massive decline in the population of many species, such as the bald eagle and the peregrine falcon.

Another nasty group of chemicals that accumulate in animal tissues are the PCBs (polychlorinated biphenyl), which were widely used around the same period as DDT. PCBs found myriad applications as inks, adhesives, flame retardants, paints and coolants for machines. Their health impacts make no pleasant reading: they are linked to cancer, lower fertility, hormonal disruptions, pain, lung damage, immune system deficiency, and more. And because they are very long-lived, hundreds of thousands of tonnes of PCB still remain in the environment, despite having been banned for decades in most countries.

An often neglected, but increasingly problematic and now ubiquitous form of pollution is caused by artificial lights. They not only affect our own sleep, disrupt our cognitive functions and our daily hormonal cycles, but also modify the behaviour of the wildlife around us. For hundreds of millions of years, species have evolved under the constant shift between day and night, light and dark. These daily variations have left manifold physiological imprints, such as the circadian clock programmed in individual tissues of mammals, including us. Today, nearly a quarter of the world's land surface is affected by artificial light at night, either directly or indirectly through skyglow. Artificial lights disorient migrating birds and sea turtles and disturb the behaviour of crickets, moths and bats. In Germany, it has been estimated that artificial lights may kill more than 60 billion insects over a single summer, which fly straight into lamps and die, or collapse after circling them for hours. Light pollution, habitat loss, pesticides, competition with invasive species, and climate change are all thought to have contributed to the massive decline in abundance and diversity of insects worldwide over recent decades, with substantial knock-on declines in the pollination of wild and cultivated plants.

Yet another form of pollution, and acknowledged even less than artificial light, is noise. This is particularly serious in the ocean, since sound travels farthest in water. Myriad life forms – from molluscs and crustaceans, to fish, dolphins, seals, turtles and whales – use sound as the sensory cue to explore the marine environment and to interact with other species and congeners. Adaptations to capture and interpret sound began some 500 million years ago with the jellyfish and have been developing gradually in other species ever since. But wildlife has been unprepared for the massive changes in the world's 'soundscape' that have occurred over the last few decades. Today, the noise of ships and boats, low-flying aeroplanes, construction work, seismic surveys, military activity, pile-driving and underwater mining for oil and gas – some of which is able to travel thousands of kilometres – is causing massive behavioural impacts, disrupting the ability of animals to move, forage, socialise, communicate, rest and respond to predators, increasing their mortality and reducing their reproductive success.

EMERGING DISEASES

All species have adaptations against certain forms of bacteria, viruses, fungi and other pathogens. These include the thick cell walls of plants and the skins of animals, which can block the way for diseases, and immune systems, which fight off infections. Sometimes, though, the species causing a disease can develop new ways of attacking, or affect a species not previously exposed to them. New variants of known viruses, caused by random mutations in their DNA, can be slightly more contagious or dangerous.

It has been estimated that bats alone may host some 5,000 different coronaviruses, while birds and mammals may host over 1.6 million unknown viruses, half of which have the potential to cross over into humans. Under normal circumstances, those viruses do not usually pose threats to us. But, as we're now disrupting the balance in ecosystems, we're also creating new opportunities for transmission. For instance, as the forests in the Ivory Coast are cut down, flying foxes – a known reservoir for viruses such as Ebola – have been forced to move to street trees in cities, where they release large quantities of faeces. It is not unthinkable that new viruses in bats or other wildlife could lead to human death rates in orders of magnitude higher than those seen in COVID-19. Tampering with the environment not only poses threats to the wild biodiversity in ecosystems, but also leads to unintended consequences that may lead us closer to disease.

While viruses and bacteria are a major concern for humans, another group of organisms – fungi – represent perhaps an even more prominent threat to some other species. One of the best-studied and most devastating cases of disease in wild animals is that of a tiny fungus with a long name, *Batrachochytrium.* Molecular studies have traced its origin to the Korean Peninsula, from where it dispersed as a result of a global expansion of the commercial trade in amphibians. Once infected by the fungus, frogs and toads suffer a skin ravage that alters their salt balance, eventually causing heart failure and death. Over a period of just two decades since its discovery in 1998, the disease led to population declines in more than 500 species, including 90 presumed extinctions. These include the two species of gastric-brooding frogs native to Queensland in eastern Australia. These remarkable creatures were unique for being the only two known frog species where the mother incubated the offspring in her stomach, until they grew into a fully developed stage capable of taking care of themselves. As a result, a human desire to expand the trade of amphibians inadvertently killed off two of their most treasured species; those whose rare characteristics we are unlikely to ever see again.

Among plants, another distantly related fungus is associated with the rapid obliteration of the iconic American chestnut in its natural range. It used to be one of the most abundant forest trees in eastern North America, but chestnut blight – a parasitic fungus – killed some three to four billion trees between its discovery in 1904 and the mid-20th century. As with the fungus affecting amphibians, the origin of this disease can be traced to East Asia, from where it travelled to North America on imported Asiatic chestnut trees.

In summary, these are the main threats to biodiversity: human encroachment on and destruction of natural habitats; unsustainable consumption and international trade in illegally captured animals and plants; a warmer climate with extreme and unpredictable weather events; increased encounters between previously disconnected and often already decimated species; multiple forms of pollution; and new diseases. This is the result: all these events are now interacting to substantially increase the risk of extinction for a large number of species, including our own. And this is not even a complete list.

Safeguarding the world's biodiversity against so many hazards may seem like a huge and complex task. But it *is* still possible, and it *is* worth it, considering the endless benefits of biodiversity, for us and the planet. Thus, we have no alternative but to do our best to save it, to keep alive this hidden universe of species on Earth. Let's see how.

PART FOUR

SAVING BIODIVERSITY

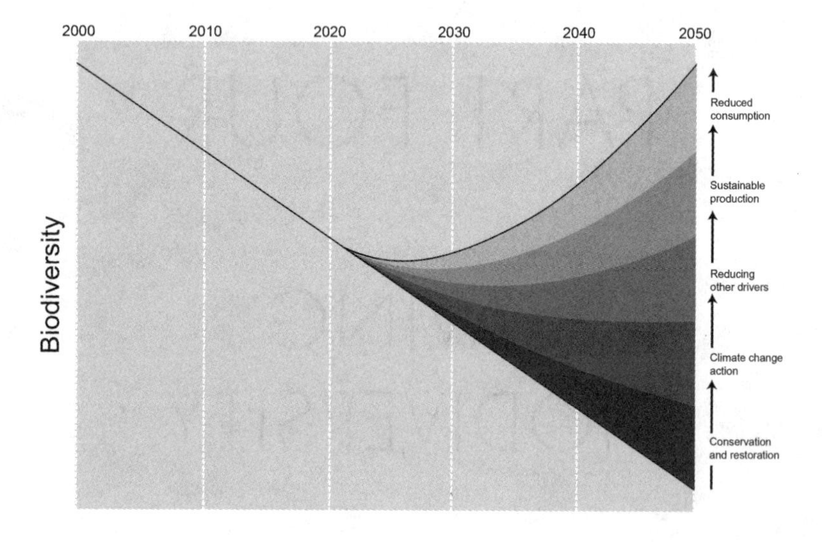

Figure 20. Bending the curve of biodiversity loss. This diagram, produced by the Secretariat of the Convention on Biological Diversity, lists the most impactful actions to protect and restore nature, which I discuss in more concrete terms in the following chapters and throughout this book. They belong to five broad categories: 1) conserving what is left and restoring what we've degraded; 2) combating climate change, through reduced emissions and increased carbon sequestration from the atmosphere; 3) reducing other drivers, by taking action against pollution, invasive species and overexploitation; 4) creating more sustainable forms of production of goods and services, particularly food; and 5) reducing consumption and waste by all of us.

Sixty-six million years ago, an asteroid landed on Earth and killed three-quarters of all species on land and at sea. It couldn't be stopped. But the tragedy resulting from the metaphorical asteroid that is on a direct collision course towards our world today can still be averted – simply because, this time, the new asteroid is us. It won't be simple, it won't be easy. But we have no choice. Just as thousands of physicists have joined forces to understand the fundamental particles that make up our universe – including dark matter – at CERN's hadron collider in Geneva, so thousands of scientists and practitioners today are working around the clock to find solutions to the biodiversity crisis. As a result, what needs to be done is well outlined, even though implementing the solutions at a scale commensurate with the challenges we face is not trivial. Clearly, we'll only save biodiversity if we radically transform the way we live our lives and interact with nature. Achieving this will require multiple concerted efforts (Fig. 20) and engagement from every segment of society, so that we all join this revolution and work together – from our top political leaders to our own children and, well, even our pets.

CHAPTER 13

LARGE-SCALE SOLUTIONS

A couple of years ago, my daughter Maria downloaded a mobile app that promised to plant a tree each time she spent a few hours doing her homework. She's well aware of the environmental crisis and was therefore proud to be making a tangible contribution to restoring forests, without even having to leave her bedroom. That got me wondering how *exactly* the app would do it: which trees would be planted, where, by whom, and would they survive and bring the intended benefits? At the same time, I knew well that companies, governments, NGOs and wealthy individuals around the world were making increasingly ambitious pledges as if they were bidding against each other: 'We'll plant a million trees!' 'We'll plant a billion trees!' 'We'll plant a trillion trees!'

I got worried. From my own experience, I knew very well that a planted forest is nothing compared to a natural one: it doesn't replace like for like. Claims by the Swedish forestry industry that 'Sweden has never had as much forest as today' were intended to mislead the public. What they referred to, in fact, were monocultures of spruce and pine, no more biodiverse than a monoculture of soy in the tropics. Natural forests, in contrast, are complex ecosystems offering habitats to a large range of species above and below ground – from worms, other invertebrates and fungi in the soil to mosses, lichens, birds, mammals and many other life forms at different layers of the canopy.

Figure 21. The common European oak, *Quercus robur*. This tree species can survive for hundreds of years and sustain about a thousand different species, including lichens, mosses, insects, birds, mammals and other organisms.

Those forests were becoming increasingly rare, while new plantations were popping up everywhere.

What is worse, in many of my trips to the tropics I had seen how plantations of exotic trees – like Australian eucalypts in Africa, and North American pines in Brazil – seemed to be doing more damage than good, despite claims (and intentions) that those plantations would help tackle climate change. On the one hand, those trees were easy to plant (in some cases even by throwing seeds from a drone), they grew fast and their timber had good commercial value. But they also sucked vast amounts of water from the watershed, decreasing supplies of drinking water for farmers and their crops. They often became invasive, spreading uncontrollably into natural forests and outcompeting native species; and they had very low levels of biodiversity, providing no food sources for native animals and modifying the soil composition in ways that made it unsuitable for most other species. The plantations also suffered more easily from pests and diseases.

Identifying a problem doesn't move us forward; we need a solution, and in this case one that could be scaled up. Inspired by my daughter's app, I worked with a small team of scientists at Kew and Botanic Gardens Conservation International, including Kate Hardwick, Alice Di Sacco, Rhian Smith and Paul Smith. Together with other experts around the world, we researched and published on the best ways to reforest habitats, based on the best available evidence. Our publication setting out 'Ten golden rules for reforestation' received huge interest from the media, reaching tens of millions of people worldwide. Thousands of organisations and individuals, many of them directly involved with tree-planting initiatives, later signed a declaration stating that they wanted reforestation projects to follow those rules. We also held a virtual conference, attended by thousands of participants from over a hundred countries, to discuss best practices and opportunities for planting trees and restoring forests. The take-home message from all those efforts is that reforestation has a huge potential to simultaneously tackle the grand challenges of biodiversity loss, climate change and poverty, but only if based on the principle of 'the right tree, in the right place, cared for in the right way'. Our guidelines also include the importance of involving local communities from the start, avoiding areas that were not previously forested, planning ahead for the effects of climate change, and carefully considering the long-term impacts on the landscape. In many cases, just letting forests regenerate by themselves is the easiest and most effective approach.

Reforestation is one among a number of promising 'nature-based solutions' to tackle global challenges, alongside other forms of restoration of important ecosystems such as mangroves, coral reefs and natural grasslands. Besides the benefits to biodiversity, climate and people's livelihoods, the restoration of terrestrial ecosystems can help against soil erosion and landslides, storm surges and inundation, salt-water intrusion, wildfires, pests and droughts, while the restoration of mangroves, corals and seagrass beds can help re-establish healthy fish populations, boost ecotourism, capture carbon and provide protection to coastal communities.

Whenever there is a choice, though, protecting the natural ecosystems that remain is always preferable to trying to restore them after their degradation. To reverse the current trend of terrestrial biodiversity loss by 2050 – the vision set out by the **Convention on Biological Diversity** – it has been estimated that some 40 per cent of the world's terrestrial areas need conservation management. Although the past decade saw a considerable increase in land area designated as protected, today only some 15 per cent of terrestrial environments are within such protected areas, meaning that we still have a long way to go. Increasing that number will require huge financial investments, as it will no longer be enough to protect cheap land in hard-to-access areas that have little economic value for agriculture or other uses, as has often been the case so far. It is crucial to focus on the biodiversity outcome rather than a mere area-based target, and to acknowledge that some countries – like Madagascar – should primarily be supported to increase the effectiveness of the existing network of protected areas, before creating new ones.

The new areas need to be well connected, through biological corridors, to enable the free movement of species for foraging, mating and coping with a changing climate. In practice, today's global network of roads – equivalent to more than 160 times the distance to the moon – presents a barrier to most animals. Some wealthy countries are now creating bridges and tunnels to allow movement, which is a good effort and should always be considered early on in any new road plans. However, it may be best not to build new roads at all. When new roads are built, they have a direct and substantial impact on natural habitats through the need to clear vegetation from the land along the planned route. But they also play a major role in facilitating degradation of the nature surrounding them. Roads provide a means to get from one place to another, but they also give people access to hunt, extract minerals and other natural products, and log forests across large areas that were previously difficult to reach. In the Amazon for instance, 94 per cent of all deforestation occurs within 5.5 kilometres (3.4 miles) of a road. Road kills can also have markedly negative impacts on certain species. In southern Nevada in the USA, for example, the population density of desert tortoises is lower within 4 kilometres (2.5 miles) of roads compared to elsewhere.

Protected areas also need to complement each other in covering as much biodiversity as possible, focusing not only on protecting high numbers of species but also evolutionary, functional, genetic and ecosystem diversity, which only partly overlap in space. It'll also be crucial for those areas to be monitored and managed effectively, rather than becoming what some people call 'paper reserves' – areas officially protected and reported as such internationally, but often indistinguishable from surrounding regions in terms of the level of biodiversity, and with no real conservation taking place.

In identifying new areas for conservation, it'll be key to focus not just on those that sustain charismatic animals – such as the habitats of gorillas and snow leopards – but also those that offer protection for diverse communities of less acclaimed groups, such as plants and fungi (Fig. 22). This is why my colleagues at Kew have worked for several years in identifying important areas for conservation around the world, with a focus on tropical countries such as Bolivia, Mozambique and Guinea-Conakry. That work

Figure 22. Protecting ecosystems and their benefits to people. The high-elevation Páramo vegetation of the Andes in South America is a prime example where plant communities not only sustain a rich diversity of life, including fungi, mosses and birds, but also function as large reservoirs of clean water for millions of people.

involves the study of satellite images, a review of previous biological inventories done in the country, additional fieldwork to survey candidate areas, and strong collaborations with local partners throughout the process. Area prioritisation can really make a difference, as when the government of Cameroon announced in 2020 logging concessions that would destroy large parts of the unique Ebo forest. By demonstrating that many unique plant species would likely go extinct, and by engaging in a campaign that even got support from Hollywood actor Leonardo DiCaprio, Kew scientists and our country partners succeeded in convincing policymakers to quickly reverse their decision and to offer long-term protection of Ebo, which brought national and international praise.

One of the most critical factors for effective conservation that we often forget is people themselves. As summarised by prominent British ecologist Georgina Mace, who sadly passed away in 2020, biodiversity conservation has shown a gradual evolution in thinking and practice: from the 1960s, it focused on 'nature for itself', followed by 'nature despite people' from the 1980s, then 'nature for people' from the 2000s. Since the 2010s, the focus has been on 'people and nature', which better recognises the importance of considering and benefitting both simultaneously.

I saw a real example of this when studying the Vohibola littoral rainforest in eastern Madagascar a few years ago, with Hélène Ralimanana and other colleagues from the Kew Madagascar Conservation Centre. We had driven for nearly ten hours through a highly degraded landscape, where we could barely spot a standing tree. But at our final destination, things looked very different: it was an incredibly diverse and beautiful fragment of the once much wider rainforest. The reason it had survived almost intact was that the owners of the lodge we stayed at had engaged intensively with the local communities, and would pay part of their income to them for keeping the forest free of logging or hunting. It was a simple model of ecotourism, and it worked, because the local people received tangible benefits towards their livelihoods. Similar models could be adopted in many places, especially if supported by local governments and systems that help promote an even distribution of tourists across a country.

Steering collective action is not a trivial task. From a policy perspective, there are two main mechanisms: incentives or punishment. Both often happen through taxes: if you lower them, or provide subsidies, you encourage people to do the 'right' thing for the environment, and vice versa. This is important, as we cannot expect every citizen to make altruistic decisions for the public good. Flying, for instance, should always be more expensive than taking the train between any two locations, given its generally much larger environmental impact (assuming railways already exist, since their construction has a large footprint).

Governments and parliaments can also create strong and fast change through new laws, bans and international agreements. After all, the fight against smoking in restaurants and other enclosed places was not won by appealing to smokers, but through smoking bans. Several international environmental bans have been successful before. In the late 1980s, the production and use of chlorofluorocarbons (CFCs) in refrigerants and propellants were phased out slowly, as the chemicals were shown to deplete the ozone layer in the stratosphere, causing a 'hole' over Antarctica 29 million square kilometres (11 million square miles) wide. The ban on whale hunting was introduced as a response to the steep decline of the populations of many whale species, including the largest animal that ever existed – the blue whale. In North America and Europe, the banning of DDT and PCBs (whose effects were discussed in the previous chapter) in the 1970s and 1980s helped save several species of eagles and seals from the brink of extinction. The more recent ban on most products containing mercury in the 2000s is also having positive effects on wildlife.

In all those cases, we've seen promising signs of recovery, giving confidence that other similar bans and regulations could lead to major benefits for species and ecosystems. It's also important to learn from other countries' experiences and follow their example. Unfortunately, diclofenac continues to be widely used as a medicament worldwide and has been recently approved for veterinary use in Spain (which contains large populations of vultures), despite the severe negative effects documented in India and other parts of South Asia. Similarly, large quantities of mercury continue to be released into the environment during mining activities in many low-income countries, despite their well-known adverse effects on the neurological and reproductive systems of animals.

One urgent situation where international legislation could lead to tangible benefits for wildlife, and also humans, would be a global ban on perfluoroalkyl substances (PFAS): a group of over 4,700 manmade 'forever chemicals' which never seem to break down. Due to their stain- and water-repellent characteristics, and the fact that they tolerate high temperatures, they are used in thousands of products, from furniture and clothing to frying pans, shoes, carpets, cosmetics, food packaging, firefighting foams, ski wax and electronics. They end up everywhere in nature, where several of them have been shown to affect the reproductive and immunological system of animals, cause hormone disruptions, increase the risk of miscarriage and probably cause several forms of cancer.

Sometimes a bit of creativity can go a long way. In the reforestation conference that I mentioned earlier, the president of Costa Rica, Carlos Alvarado Quesada, shared the remarkable history of how his country went from having one of the highest deforestation rates in Latin America in the 1980s, to becoming one of the greenest and most sustainably managed countries in the world, where biodiversity is a matter of national pride and income. The way they flipped things around was an act of genius: they simply decided to use fuel taxes to pay landowners to stop cutting down their forests. Those taxes were enough to give them US $42 per hectare (100 x 100 metres), every year, a value that gradually increased to over US $80. It may not sound like a lot, but it was just enough to make it more profitable to keep a forest intact than to convert it into farmland. People could still pursue other non-destructive forest-based businesses, like organising birdwatching excursions for tourists; or they could simply enjoy life and go for a walk under the rich canopy with singing birds, or sit under a tree and play the guitar!

Global change to save biodiversity must be supported by legal frameworks that encourage positive action *and* hold nations and their leaders accountable when they are not doing enough. A key concept here is to consider the large-scale destruction of ecosystems, with far-reaching consequences for people and biodiversity, as an international crime. Calls for recognising **ecocide**, the environmental equivalent of genocide, have been made by conservationists for decades, but over the last few years the concept has been gaining momentum – growing in popularity and receiving increased media attention. Ecocide was almost included in the Rome Statute in the 1990s, the treaty that established the International Criminal Court in The Hague, but was removed at the last minute as a result of pressure from the Netherlands, France and the United Kingdom. In June 2021, following a request from Swedish members of Parliament Rebecka Le Moine and Magnus Manhammar, a new legal definition of ecocide was proposed, to mean '*unlawful or wanton acts committed with knowledge that there is a substantial likelihood of severe and either widespread or long-term damage to the environment being caused by those acts*'. Writing with activist Pella Thiel, we have articulated why this proposal should be taken seriously and implemented into national and international laws, alongside other measures discussed here, to avoid individual governmental and corporate leaders creating environmental damage that could be perfectly easily avoided through the implementation of science-based practices and international cooperation.

Protecting remaining ecosystems and restoring degraded ones are crucial steps in moving forward, but will fall short unless we simultaneously tackle the most important underlying driver of biodiversity loss: the need for food. As the global human population doubled from 3.9 to 7.8 billion over the course of just five decades, and is expected to continue increasing at least until the middle of this century, we must find ways to reduce the pressure that food production has on marine and terrestrial ecosystems. It has been estimated that we will need to produce more food in the next 50 years than we have ever produced in human history, which will require a deep transformation in the way we produce and consume it.

On land, a widely advocated solution is to intensify farming as much as possible, by using increasingly larger and specialised machines and sometimes involving genetically modified crops. This is already the standard in many middle- to high-income countries, for example in the vast wheat crops of Canada, the USA and Western Australia. But such a system isn't suitable everywhere; it is expensive to set up, excludes wider participation of local communities, is most suitable for flat areas, relies on strong pesticides, and excludes most birds, insects and other wildlife from the cultivated land. While intensive farming has the potential to reduce the need to expand farmlands into previously natural ecosystems, in practice this has seldom happened so far – as seen by the continuous expansion of soya fields into the Amazon, to meet an ever-increasing global demand.

A better alternative, especially in many low-income and biodiverse regions, may be to promote traditional practices by small-scale farmers. Families often work together to grow a variety of crops, depending on the local environmental conditions, such as soil type and microclimate. Their produce is then traded within and between communities – work often led by women. Although variants of this system have been carried out for millennia around the world, over the last century many locally used crops that aren't traded internationally (orphan crops) and wild species of plants related to domesticated crops (crop wild relatives) have become increasingly neglected in favour of a handful of dominant crops, with a consequent loss of nutrient levels in our diets. There's a great opportunity to diversify those crops, increasing, what some call, agrobiodiversity. Scientists are working on many projects to support such transitions and advise on suitable crops that will not only do well in today's climate but also be more resilient to tomorrow's challenges in the face of climate change. Some of these, like yams in Ethiopia and Madagascar, which have been a focus of research for my colleague at Kew, Paul Wilkin, Sebsebe Demissew in Addis Ababa, Ethiopia, and their collaborators, grow best in the shade offered by forests. Their cultivation (agroforestry) reduces the need for felling trees, which helps protect the local biodiversity and all the benefits forests bring – such as clean water supply, air cooling and prevention of erosion and flooding. Increasing agrobiodiversity, particularly alongside programmes supporting education and family planning, can therefore bring many positive outcomes, from reducing poverty to improving health, food security and biodiversity protection. I think this should be the focus of many programmes supported by local and international aid funds.

A final but crucial point about food is the imperative to reduce waste. The numbers speak for themselves: one-third of the world's food is wasted, every single year. This would be enough to feed the 815 million hungry people in the world, four times over. Total food waste is a combination of many factors, from the long journey of food from farm to table, to the exaggerated pickiness of customers for perfect fruits and vegetables. Although individual consumers have a major role to play, so do many other segments of society: schools, companies, hotels, restaurants, supermarkets, governments. And this is beginning to happen. In France, it's illegal for supermarkets to discard unused food, which needs to be donated. Some countries, like Denmark and Germany, have set up businesses that sell produce past their 'sell by' date at a reduced price. Public campaigns are encouraging people to buy local produce, while special gelatin tags have been invented to tell when a food has really gone bad. It's now time to scale up some of those great initiatives and support further innovation in this area.

Waste reduction, alongside composting and recycling, contributes not only to reducing the demand on land but also to reducing the leakage of nutrients, pesticides and other pollutants into the environment, which pose further threats to biodiversity and people. This is just one example of the advantages of making all manufacturing circular – where nothing is lost along the way and is used over and over again. **Circular production** is just one way of mimicking nature, where elements are constantly recycled – such as trees decomposing and releasing nitrogen and phosphorus into the soil, only for them to be picked up again by the next tree growing in its place.

The private sector is a key player in transforming societies towards sustainability. Companies hold the power, innovation and resources needed to transform their products and services to meet the increasing environmental demands of customers. Efforts such as the United Nations Global Compact and the Sustainable Markets Initiative are now gathering momentum by bringing together the leaders of major companies to ensure that every investment they make moves them towards a greener future.

Searching for inspiration in nature – the field of **biomimetics** – is another example where companies can unleash the untold uses of biodiversity for solving many of our problems and improving wellbeing. Examples include the design of Japan's fast-speed Shinkansen trains taking inspiration from the beaks of kingfishers to become faster and quieter, particularly as they pass through tunnels where air resistance slows them down; the Velcro straps invented by Swiss engineer George de Mestral after he removed burrs (hooked seeds or fruits) from his dog, and decided to have a closer look at them; and the shopping centre in Zimbabwe that mimics a termite mound, and keeps a constant cool temperature using only 10 per cent of the energy needed by a similar-sized conventional building.

Ultimately, nations around the world need to tackle not only the direct drivers of biodiversity loss – habitat degradation on land and in the sea, exploitation, climate change, pollution, invasive species – but also the indirect drivers, such as population growth, poverty, conflicts and epidemics. The synergy of drivers clearly shows the strong and complex links and interdependence between human and natural systems.

Some could argue that halting biodiversity loss would just be too expensive in monetary terms, considering the opportunity costs, such as not being able to cultivate crops on fertile land, or having to divert public funding towards habitat restoration rather than healthcare, education or the military. In fact, this is due to the way that natural assets have been valued – or indeed, *not valued* – historically. As economist Partha Dasgupta has shown, we have systematically used natural ecosystems and their species to build our societies and support our consumption without paying for their extraction or replacement. In a period of just over two decades (between 1992 and 2014), global investments led to an increase of almost 100 per cent in **produced capital**, such as buildings, roads and machines. In contrast, the stock of **natural capital**, such as forests, decreased by nearly 40 per cent. It may not sound like an investment to protect a piece of forest, but it is, since this allows it to grow and develop its capacity to capture carbon, stabilise the soil, reduce flooding and provide pollination, clean water, building material, shade and many other services and goods. Even in current monetary systems, it is clear that investing in biodiversity protection now rather than later will be much cheaper: in fact, a report by the Natural History Museum in London estimated that waiting a decade would double the costs and lead to the loss of many more species.

Bhutan and, more recently, New Zealand have shown the way: they are shifting their priorities and public investments to achieve wider environmental and social benefits. Those two countries have replaced the standard pursuit of economic growth prioritised by nearly all nations, often measured by a single metric (the Gross Domestic Product), with another model, using instead more inclusive measures of wealth based on the quality of natural ecosystems and the wellbeing of people. They are also embracing diverse visions of what it means to have a 'good' life that is less linked to monetary assets and produced capital. Their examples show that a transformation of the financial system is not only achievable, but also essential for a fairer distribution of the world's dwindling resources. It is now time for all other nations to place proper value on natural capital and to internalise the costs of ecosystem degradation, biodiversity loss, pollution and climate change.

Challenging economic growth and including environmental impacts in economics may sound like a new, revolutionary idea, but it is not. In 1972 in the USA, economist William Nordhaus wrote a seminal paper discussing this very issue and referring to a quote by ecologist Paul Ehrlich that '*We must acquire a life style which has as its goal maximum freedom and happiness for the individual, not a maximum Gross National Product*'. In the decades that followed, Nordhaus went on to develop a series of influential economic models that would consider the true environmental impact of growth, which eventually won him a Nobel Prize in Economics in 2018.

Nature is essential for sustaining human wellbeing for current and future generations. As widely advocated by Argentinian ecologist Sandra Díaz and her colleagues, who have reviewed a large number of studies from across the world, there is ample scientific evidence that a healthy planet also means healthy people. The solutions I outlined here represent just a few of the areas in which societies around the world need to focus. The key to transformation is large-scale initiatives delivered at national and global levels. At the same time, there are many things we can do as individuals to support biodiversity and reduce our own environmental impacts, delivered through the actions and choices that each of us make in our daily lives. These actions reinforce and amplify each other: a shift in individual action can both enable and compel governments to go further, which in turn increases public uptake of the changes (as seen with the ban on indoor smoking) and forces companies and businesses to adapt at pace, preventing public recalcitrance and again encouraging governments to go further ... in an escalating loop of positive change. Combined, these two sets of changes, both on global and personal scales, are the most powerful tools for protecting the world's biodiversity.

CHAPTER 14

WHAT CAN WE DO?

'If you think you are too small to make a difference, you haven't spent the night with a mosquito,' says an African proverb. The threats facing the world's biodiversity may seem daunting, but we can all play a critical role, which combined will lead to massive positive change

There isn't a single, one-size-fits-all way for individuals to contribute. We all have different roles in our societies, personal networks, job situations and economic capabilities. While some people own gardens or land, and have a direct influence on the species inhabiting them, many others live in flats with a balcony at most. If this is you, however, you are still able to make a similar – or even a bigger – difference through your consumer choices and actions. I wish I could say that there's no point in feeling bad about not doing much or enough, and that any action is better than no action. But the truth is that the only way to stop and reverse the loss of the world's incredible biodiversity is for each of us to change our lifestyle radically and substantially, and to do that now.

The great news is that if each of us *does* substantially reduce our environmental footprint, and we influence others to do the same, the combined effect will be transformational. Your actions, values and words can inspire and multiply, not least by publicly supporting positive actions and behaviours from others. In addition, there are lots of win-wins, and very few sacrifices: what is good for biodiversity and for the climate is nearly always also good for our health and wellbeing and our pocket.

Below I list key areas of change that we all can make. These include actions that directly reduce our negative impact on biodiversity, as well as indirect benefits, which are beneficial to the environment and to combating climate change more broadly. Some items will be easier and faster to adopt than others; as long as we make progress year on year, we'll be moving in the right direction. These recommendations are by no means exhaustive; rather, this is mainly a personal list of things I've been trying to pursue together with my family.

FOOD

With food production being a main driver of biodiversity loss, this is where we can all start to take positive action at a personal level. Growing up in Brazil, my plate looked almost identical every day: rice, beans, some salad and, because my family could afford it, a slice of meat. There and almost everywhere else, the consumption of meat – in particular beef, pork and chicken – has the most damaging effects on terrestrial biodiversity. This is because animals require substantially more energy, land and water to grow when compared to plants, which can provide us with proteins and other nutrients directly. For instance, to produce a kilogram of beef requires on average more than 15,000 litres of water, compared to 255 litres for potatoes, and this is water that is often diverted away from biodiverse-rich wetlands and river systems. Worldwide, more than 40 per cent of all wheat, rye, oat and maize production is fed to livestock rather than to humans directly, along with 250 million tons of soybeans and other oilseeds each year.

I know it's not easy to change one's diet. Food is a major part of our cultures, indeed our identities, and we all remember with fondness dishes we ate as children. Memories associated with particular smells and tastes are stored in our brains far more efficiently than just images. Anyone who has tried to lose weight by following a new diet, for instance, knows just how difficult it is to maintain it over a longer period, and how easy it is to fall back into previous habits. In many countries, the consumption of meat, in particular, is linked to social and economic status, and barbecues are a matter of national pride and social exchange in almost every country, from Mexico's *barbacoa* to South Africa's *braai* and Japan's *yakitori*. Yet, changing what we eat is precisely what we need to do, especially in parts of the world with high per capita consumption of meat. Already today, some four billion people live primarily on a plant-based diet, particularly in India and Africa, but meat consumption is unsustainably high in most other regions, and has increased rapidly in the majority of countries worldwide.

Besides its generally high price, meat also takes a toll on our health. People who eat a lot of red and processed meat suffer a higher mortality rate from cardiovascular diseases, and there is a strong association between a high intake of processed meat and colorectal cancer. Beyond our individual health benefits, reducing meat consumption will also reduce one of our most concerning public health problems: antibiotic resistance. This is an unsettling trend, where bacteria responsible for previously easily treatable diseases change their DNA in response to widespread use (and misuse) of antibiotics, and thereby no longer respond to treatment. Each year, China alone uses over 80,000 tons of antibiotics on livestock to prevent disease spreading between animals kept in tightly enclosed spaces, with some 50,000 tons ending up in the soil and rivers, affecting wildlife and people, particularly children. If we don't stop our overconsumption of meat, not only will our health suffer directly now, but so will the health of those around us, and others across the world, with long-lasting effects.

➤ ***Eat less meat, or none at all.***

As articulated above, cutting out meat from our diets and replacing it with sustainably sourced alternatives will have one of the biggest positive effects on biodiversity, climate and the environment, by reducing the pressure on land and natural resources, including in my home country Brazil – the largest producer of soybeans for cattle feed. If you do want to eat meat occasionally, choose organically grown, locally produced products. Choose fish belonging to species with sustainable population sizes, preferably caught with baits, trawls or nets that reduce by-catches (releasing fish that have not yet reached reproductive age, and larger animals like dolphins and seals). Skipping meat, or reducing its consumption, also means reducing the single most important source of methane, which is released in large quantities by ruminating livestock and constitutes the second most significant greenhouse gas after carbon dioxide (in fact it is even more potent, but lasts for a shorter time and is present at much lower concentrations than carbon dioxide). By shifting to a plant-based diet, each of us saves nearly a ton of greenhouse gases per year. Dairy products such as cheese, yoghurt and butter can also have a big environmental impact, depending on how the animals were fed. One alternative option is to eat insects – something regarded as disgusting and primitive by some, but insects form part of the traditional diet of over two billion people around the world. More than 1,900 insect species are regularly consumed by humans across the planet, often with minor environmental consequences, especially as some can be grown on organic biowaste – which is preferable to capturing them in the wild, since many species have declined dramatically around the world. If an insect diet doesn't sound appealing to you, you're not alone, but if you try one day, you might actually be surprised – I certainly was – by how yummy fried grasshoppers, ants and beetles can be! Lab-grown meat is yet another area attracting interest, which brings environmental benefits as compared to traditional farming. However, it sustains people's meat-eating habits, which might counteract the rapid and widespread transition away from animal products that nature now needs.

➤ ***Eat more fruit and vegetables.***

Luckily, there's an enormous, and still poorly explored, diversity in the non-animal kingdom to replace meat with. My colleagues Tiziana Ulian, Mauricio Diazgranados, Sam Pironon and others at Kew have identified over 7,000 plant species that are used as food sources around the world and are at the same time nutritious, robust enough for a changing climate, and at low risk of extinction. Although you may never have heard about many of them, some are consumed locally by millions of people. Like the morama bean, a southern African legume with seeds that taste like cashew nuts when roasted, and can be boiled or ground to a powder to make porridge or a cocoa-like drink. Or the pandan tree, which is a drought-tolerant genus that grows in coastal lowlands from Hawaii to the Philippines and produces a fruit that can be eaten raw or cooked. Which options you have will depend on where you live and what is available in markets around you. Give preference to locally available species and varieties, and choose by season. I know avocados are delicious, but if they don't grow where you live it means long transportations of a crop that already has a large environmental impact, from intensive water usage in Chile to blocking the natural

Figure 23. Wonky fruit and vegetables. Human resistance to this natural variation means that these are almost invariably thrown away at the end of the day, if they ever make it to the supermarket shelves. This is part of the reason why so much food is wasted, despite so many people still going hungry and the pressure that agriculture poses on biodiversity.

migration routes of elephants in Kenya. When shopping, look for wonky fruits and vegetables (Fig. 23), rather than perfectly cylindrical and spotless items (in the United Kingdom for instance, 40 per cent of potatoes, apples and onions are tossed in the bin, and 25 per cent of carrots are rejected because of cosmetic imperfections). Bring your own fabric bags to the supermarket and avoid products with plastic if possible.

➤ ***Expand your diet.***

Fungi and algae are great supplements to a plant-based diet, adding important nutrients to your plate – and they can be grown with minimal environmental ramifications. Oyster mushrooms (Fig. 24), for example, are a great source of B vitamins, phosphorus, potassium, iron, copper and several other minerals, and they grow superbly on by-products of the food industry, such as leftovers from beer production. I once tried to grow mushrooms in my cellar, and I was surprised by how easy it was with a simple kit and minimal space requirements. In the ocean, algae can be cultivated without needing to compete for space with crops on land, and without the need for added nutrients or pesticides. They provide proteins, vitamins, minerals, antioxidants, sugars and lipids in our diets. I have visited Japan a couple of times for conferences, and one of my fondest memories from my visits was seeing how often algae and fungi are integral to so many local dishes. The potential menu is huge, with many thousand species of large edible mushrooms known. However, never eat a wild mushroom

Figure 24. Oyster mushrooms. Fungi are tasty and nutritious components of a healthy diet, and their production can be more sustainable than many other types of food.

you do not fully recognise and know is edible, or it might be the last time you do!

➤ ***Learn to cook from raw ingredients and trust your senses.***

Cooking is a bit like learning to play a musical instrument: it takes some time and effort, but soon you start to reap the rewards that come with practice. To start you off, there are lots of great recipes focused on non-animal diets on the internet and in cookbooks, but don't be too strict in following them – even if you swap several of the suggested ingredients with those you happen to have in the fridge, the result can be equally good or even better, and minimises waste. Learn to cook from the real stuff – raw rather than processed ingredients – as fresh produce provides a good base for most cooking, and is less often packaged with the unnecessary plastic characteristic of ready meals. When it comes to throwing away food, trust your mammalian nose more than a label – it has evolved for millions of years into a highly sophisticated tool to tell you whether something has truly gone bad, and 'best before' dates are just a guide.

AT HOME

Choosing our food carefully is crucial, but to minimise our environmental footprint we must think about everything else we buy, starting with what we bring into our homes.

➤ ***Avoid conventional cotton.***
The clothing industry is heavily focused on cotton, a plant that requires vast amounts of fresh water to grow, often in places where the use of water competes with the needs of people and natural ecosystems like wetlands and rivers. Conventional cotton plantations are also heavily sprayed with pesticides, which in turn leak into rivers and affect local communities as well as wildlife. Low-income countries such as India and Bangladesh are particularly affected. Organic cotton is a better alternative but requires just as much water; so look for environmentally kinder sources of plant fibres, such as hemp, flax, ramie and bamboo. To produce one kilogram of hemp (Fig. 25), for instance, 400 litres of water are used, compared to some 10,000 litres for cotton. Certain animal-based natural products such as wool can also be sustainably produced and are long lasting. After hemp, wool consumes less energy and has a smaller carbon footprint than most other textile fibres, partly because sheep can be raised on non-arable lands and rough terrains, meaning that they do not require the conversion of forests and other ecosystems into pastures as is often done for cattle (a process which releases vast amounts of carbon into the atmosphere). Sheep fleeces regrow every year after shearing, making wool a renewable material. Likewise, leather that is a by-product of organic farming is a great use of resources. Just avoid items that have been heavily processed chemically with dyes and finishing products. Clothes made out of multiple materials can be challenging to recycle, but not necessarily worse for

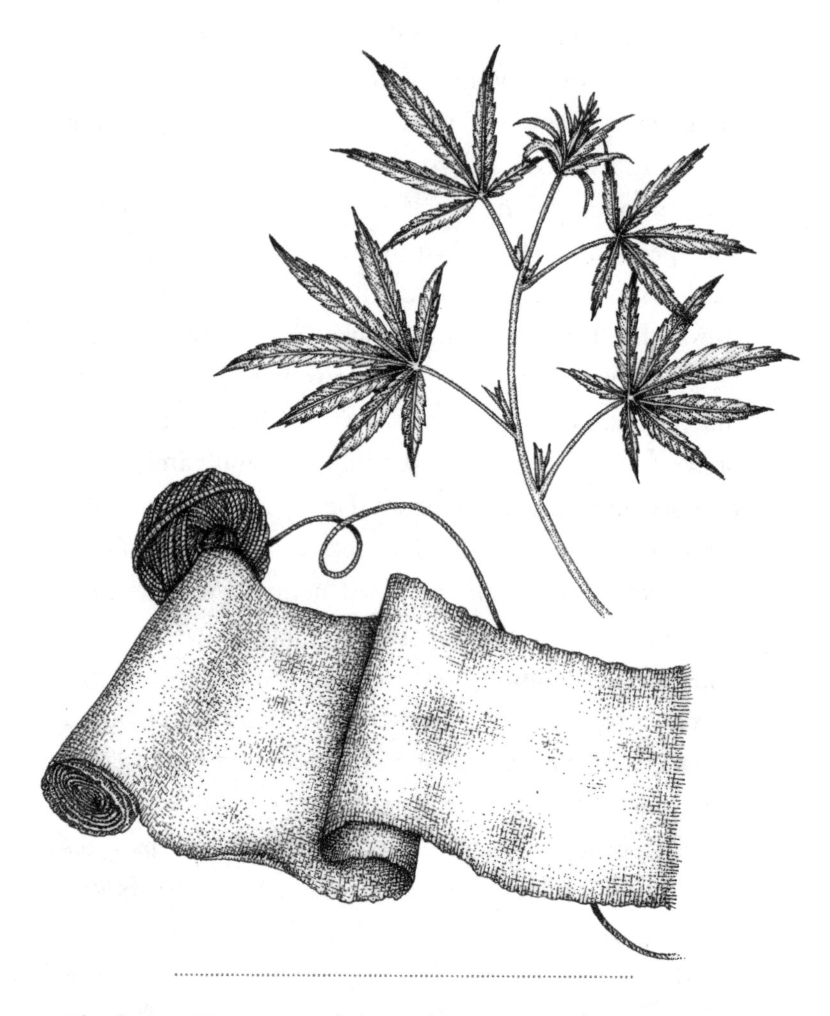

Figure 25. Hemp – a plant making the clothing industry more sustainable. Despite its main association with its use as a recreational drug, and more recently as a medicine, hemp is a highly versatile plant that has been cultivated for nearly 5,000 years as a source of fibre. Its production is far more sustainable than that of cotton.

the environment than synthetic fleece, which on the plus side can be made out of used plastic bottles but on the other hand releases microplastics when washed, ending up everywhere in the food web with worrying consequences to wildlife and us. New materials are regularly appearing in the market, developed through emerging green technologies. Fabrics under the names rayon, modal and lyocell, for instance, are all made out of cellulose fibres present in any plant, including by-products from the timber industry.

➤ ***Buy less stuff.***
There's no other choice: to reduce the pressure on natural resources and ecosystems, we must substantially decrease all forms of consumption. This will not only reduce the demand on the specific products we buy, but also the associated items and services, such as their packaging, transportation, storage and disposal (including their contribution to global waste and environmental pollution). We have to stop buying and giving away new products all the time, which means revisiting our daily habits as well as occasions such as Christmas and birthdays. If you want to give a gift, consider an experience rather than a physical gift: an exhibit, theatre, course or massage. It's also fun to swap things with others, like clothes, books and plants, at swapping events and street markets organised in many cities. You can hire special clothes for fancy dinners or gear for special sports – often many times before the hire cost exceeds that of buying them. Buying second-hand comes next: good furniture can be used for decades,

if not centuries, and the market for used items is extensive, comprising nearly anything you can think of – from kitchen gear to hats, mobile phones and specialist tools, often at very low prices. There are plenty of online trading sites, almost regardless of where you live. Jewellery is a hidden villain, as the mining and extraction of precious metals and jewels, such as gold and diamonds, have a huge environmental impact (recycling, re-using and repurposing are better alternatives). Be wary of 'free' marketing items – from paper brochures and keyrings to bags, pens and plastic toys; you *don't* need any of that (and neither should your company give away such things).

➤ ***Watch out for furniture.***

With about a third of all tree species currently under threat, furniture made from wood is a key area for improvement, given its strong links to deforestation and unsustainable selective logging. As I discussed in Chapter 10, there is much more illegal wood out in the market than we would like to think. The few forest certifications available are a good step forward, but have been criticised for not being as rigorous and effective as they should be. If you must buy new, consider what type of wood you really need. Avoid the most resistant tropical woods (hardwoods such as teak or mahogany) for indoor use; and even outdoor furniture made from oak and other common species can withstand age and weather if properly cared for – finished for instance with natural oil. If tropical hardwoods are indeed required, such as for fine pieces of woodworking and musical instruments, carefully check documentation to ensure that the wood was sustainably cultivated rather than harvested from the wild, and avoid threatened species, such as various species of rosewood and ebony.

➤ ***Leave wild species in their habitats.***

Wood is just one among a large number of natural materials to think selectively about before bringing home. Others, such as souvenirs made out of shells, corals or other wildlife, should never be bought, unless truly sustainable, like certain artefacts made by local communities from renewable sources – plant baskets for example (but if in doubt, best to skip it). Avoid buying unusual plants and pets: the black market is huge. Anyone with a social media account can start selling illegally collected species – from orchids and cacti to tortoises and chameleons – wiping out entire populations and even entire species, particularly those that are only found in a few places or in one small region. This problem worsened during the restrictions resulting from COVID-19, as people who were previously dependent on tourism or other activities were severely affected by the pandemic financially and had to find new ways to make a living. As consumers and citizens, we should try to support them in other, more sustainable ways, such as buying organically produced crops from that region.

➤ ***Invest in quality.***

Whether buying second-hand or new, choose things that will last. It's easy to fall into the mistake of buying the cheapest things, such as items on sale, even though they may not be exactly what we need. While TVs from the 1960s or a telephone from the 1980s could last for decades, this can't be said for today's equipment and electronics – something we as consumers should not accept. We must demand better, not least because the many metals and other components they contain are derived from environmentally damaging mining activities. Choose brands based on guarantees for their longevity, and lobby companies for longer-lasting products if you can. Also consider repairing rather than replacing items, and choose brands that promote this.

➤ ***Detox your home.***

The use of cleaning chemicals is nearly always unnecessary, and a contributing factor to the pollution of watersheds and our oceans, affecting their unique biodiversity. You may be surprised by how far you can go with using common soap or detergent for general cleaning, and vinegar in the bathroom and kitchen, including tile surfaces, sinks, toilets, bath tubs and showers; check out the internet for homemade recipes. Cosmetics and hygiene products are another important source of dangerous chemicals, from nail polish and lipsticks containing microplastics, to shampoos, soaps and toothpaste containing at least one among tens of damaging antibacterial and antifungal substances, such as triclosan. Whether washed down the sink or disposed of in the bin, these toxins are ending up everywhere in the environment, poisoning rivers and oceans, with documented effects on coral reefs, algae, turtles and fish, and contributing to bacterial resistance. Several also contain unsustainable ingredients, such as non-certified palm oil. If you don't know what an ingredient is or the impact its production has on the environment, search for it on the internet. Besides looking for environmentally friendly options for cosmetics, we should challenge the factors that drive their exaggerated use, such as the constant plugging of cleaning products and make-up by companies, and the social requirement of daily showers or baths. Also, if you do choose to use them, always dispose of left-over toxic products and medicines appropriately, for example in a recycling station.

➤ ***Buy cleaner energy.***

Households account for a substantial proportion of the world's energy consumption and carbon emissions. If you can, make sure that you're only using renewable, environmentally friendly energy – especially from solar, wind and water supplies. These are all far better than relying on the burning of coal or oil. Besides the impact of fossil fuels on climate change, their production in many regions has a direct influence on ecosystems, in particular when it comes from mining and oil extraction, which occasionally cause sea spills of catastrophic levels, with long-lasting effects on wildlife. Nuclear power is dubious: it requires a small land surface to produce the electricity, but still requires initial mining of the ore, produces radioactive waste that remains a hazard for thousands of years and is occasionally released into the environment by accidents caused by natural disasters or human error. If your provider isn't offering an energy source that you feel confident about, choose another company. Remember, though, that when it comes to energy production, no source is entirely free from problems, even those considered to be 'clean'. Wind turbines are resource-demanding to produce and have been blamed for deteriorating landscape views, making noise, and killing birds and bats (although current evidence doesn't indicate this is a major issue); solar cells require vast amounts of energy in their production, but outweigh the cons if used in the right region; and hydroelectric stations have limited environmental impact when they are up and running, but can have massive negative effects on surrounding ecosystems and watersheds, permanently hindering the free movement of animals such as migratory fish. Research and compare your local options. Despite these issues, on balance clean energy is always the better option, and it is becoming increasingly affordable. In fact, the International Energy Agency has established that solar is now the cheapest energy source in human history.

➤ ***Lower your household's energy and water consumption.*** The considerations above mean that even if using the most environmentally sustainable energy provider available where you live, it's still always critical to lower energy consumption. There are many things you can do to achieve this. You can improve the insulation of your home and, if you require heating, lower the indoor temperature a couple of degrees (just put a jumper on). You can take quicker showers with a water-saving hose, avoid bathing, and cover pots with a lid when cooking (which can cut energy use eight-fold). You can change conventional bulbs to LED, switch stand-by electronics completely off, and switch off the lights after leaving a room. You can also hang washed clothes for drying rather than using a tumble dryer. If a machine breaks, get it fixed rather than buying a new one if you can, or else make sure any new acquisition is as energy-efficient as possible, as their consumption can vary greatly (for instance, electric ovens are at least twice as efficient as those that use gas). And while saving water is particularly critical in some regions of the world with constant or seasonal water shortages, it brings several other benefits, such as reducing the energy needed to clean and warm it, and decreasing the pressure on exploiting groundwater and other sources that are also used by wildlife.

➤ ***Think twice about pets.***
Cats and dogs are wonderful companions – they reduce our stress levels and become true family members. Pet ownership is skyrocketing, with a new boost during the COVID-19 pandemic. But they are still animals with innate instincts, and can come with a considerable environmental footprint. Cats in particular will chase birds, rodents and other wild species even if properly fed at home. In the USA alone, they kill up to 4 billion birds and 22 billion mammals every year. Worst are free-ranging cats that have either escaped or have been abandoned. Islands, and their unique biota, are particularly affected, exemplified by the

devastating effects reported on Hawaii (see Chapter 7). If you're not willing to keep your cats indoors, at least put a bell on their collars to warn potential prey (it'll help a little but not much). Just as you look to your own food consumption, you need to consider theirs. The food consumed annually by a medium-sized dog, like a Labrador, is responsible for about twice the carbon emissions of a big car driven for 10,000 kilometres (6,000 miles). Consider also the choice of pet. While the food consumed by a large dog requires some 1.1 hectares (2.7 acres) of land to grow, this number drops to 0.014 hectares for a hamster. Vegetarian and vegan diets seem to work poorly for dogs and cats, but look out for new brands of insect-based pet food to reduce their carbon footprint. On the plus side, pet owners tend to exercise more, and travel and fly less. These are all important considerations to make before getting a pet for the first time, or an additional pet, or when your pet dies and you think about replacing it with another one. As a kid I grew up with dogs and I know how emotionally connected we can get to them, but perhaps it's time we learn to appreciate the animals that are free to roam in nature more than those we lock up at home.

OUR BACKYARD

Domestic gardens are a great place where simple actions can make a concrete difference to local and regional biodiversity. Today, more than half the world's population live in urban areas, a number that will increase by more than two-thirds by 2050. Taken together, gardens occupy large proportions of most urban areas – about a quarter of an average city in the United Kingdom, for instance – so we need to make the best out of them. Since we've already transformed so much of the planet's land surface, leaving little room for other species, our own backyard can with relatively modest efforts and a few active steps become a haven for wildlife, together with urban parks and other microhabitats along roads, on roundabouts and in public areas. This will bring strong benefits for supporting and increasing biodiversity, facilitating the natural movement of species across urban landscapes, reducing air and noise pollution in cities, and providing tangible benefits to our own mental health and wellbeing.

➤ ***Ditch the lawn.***
Lawns are a thirsty, time-intensive, unproductive and unnecessary use of land. Instead, let wild plant species develop into a meadow; you can often buy seeds to start off, and they will come back by themselves, but make sure they are from native species locally sourced, rather than exotic ones or those with dubious genetic provenance. If you like lawns, then consider letting the sward grow a little taller so that short sward flowering herbs and fungi can thrive. And never use fertiliser or weedkiller. If you have plenty of space, let the forest come in, either naturally (through natural regeneration from seeds) or by planting tree saplings of native trees; then you'll also attract and enjoy the many associated species and other benefits that

Figure 26. Biodiversity in our backyard. Our gardens can become havens for many species, with just a little help. A pile of logs provides a home for hedgehogs and amphibians, while nettles and other wild plants provide food and shelter for butterflies. There are many ways we can all help species in our backyard; find out what suits your local conditions and region.

trees bring. Do less weeding: it's fun to see what native plants come up by themselves, and a few corners of nettles or other native plants provide room and food for many butterflies. Cultivate species that are good sources of nectar for insects – it's best to ask for local advice. Place a pile of branches and leaves to welcome hedgehogs and small rodents and insects (Fig. 26). Leave out a few tree logs to naturally decay, and over time they will start hosting fungi, mosses and insects that depend on that substrate, which has become increasingly rare due to the obsession with cleanliness in parks and gardens.

➤ ***Create homes for others.***

Set up nests for solitary bees and wasps, birds and bats: these are easy and fun to make yourself, but are also available to purchase on the internet and from some garden shops. Feeding birds is enjoyable for us, and it increases their chances of survival throughout the year, especially through the cold or dry season. I usually mix different kinds of seeds and nuts, since birds need a varied diet too and have different requirements and preferences depending on the species. The feeders may also bring occasional visits from mammals, like squirrels. In the tropics, sliced up papayas, bananas and other fruits are sure to be appreciated. If you're feeling ambitious and have enough space, build a pond: it will invariably attract a very different diversity of insects, from water bugs to dragonflies, and hopefully salamanders, frogs and more, depending on which part of the world you're in. I built one recently and it was one of the most rewarding projects in our garden, with lots of wildlife coming in and always something for the family to look at.

➤ ***Set up compost bins.***

One of the first things I learned after getting my first job at a botanical garden is that 'the compost is the heart of any garden'. Over the years, I've learned that this couldn't be more true: it's an amazing asset, converting the entire plant-based food leftovers of a household into high-quality, nutritious soil – you'll never need to buy fertilisers (which require vast amounts of energy to be produced and contribute to breaking the nutrient balance in rivers and lakes, affecting their wildlife negatively). We have two large compost heaps at home for warm compost, which are tight enough to avoid rats. It's become almost a hobby for me to manage and care for them. They'll also take most of the branches pruned from our hedge and trees, after processing them in a garden shredder.

➤ ***Switch off outdoor lights.***

Have you ever thought about how difficult it has become to see the stars in almost any city nowadays? It feels terrible to realise that, despite the fact that some 5,000 stars are visible to the average naked eye, in reality fewer than a dozen can be seen from a typical city. Many children now grow up without ever seeing the Milky Way, or experiencing the thrills of a meteor shower, other than in movies or photos. And as we saw in Chapter 12, artificial light pollution is not only a nuisance but a huge, increasing yet underappreciated problem for biodiversity and ourselves. Night lights disrupt the circadian rhythm of animals, plants and ourselves. It's deeply damaging to our environment, so if you can avoid it, don't contribute to it!

TRANSPORTATION

Globalisation has brought all of us closer together. It's never been easier and more affordable to travel long and short distances, thanks to an explosion in the number of competing air companies, attractive payment plans for cars, government investments in roads, and more. The use of fossil fuels and emission of greenhouse gases are the key problems, but there are many more. The release of small particles from exhaust fumes and road wear, together with toxic nitrogen oxides, affects the respiratory systems of humans and other animals and can be linked to premature death. Artificial lights along roads decrease the risk of fatal accidents but greatly contribute to light pollution. Direct collisions with vehicles are also a major threat to many species, from endangered frogs to mammals, reptiles and birds. Heavy traffic leads to wear and tear of roads, requiring more asphalt and concrete to be produced, which releases high quantities of harmful gases. New roads are also key drivers of deforestation and land-use change. In all, fast and cheap transportation has made our lives a lot easier, but at a large detrimental cost to our planet. Luckily, it is within our power to do something about it.

➤ ***Work from home more often, if you can.***
The COVID-19 pandemic brought a hiatus in our frenetic travelling, but we must reduce the negative environmental impacts of transportation on a more permanent basis. Conveniently, many people quickly became experts in video-meetings. If companies encourage staff to work from home for at least one day a week – much-needed time for many people to get on with emails, or focus on particular tasks without interruptions – this would greatly reduce overall levels of transportation, and also help us all to have a healthier work-life balance.

➤ ***Walk or cycle.***
These are always the best transportation options, from both environmental and health perspectives, and, according to statistics from around the world, should be feasible for most people globally. In China, for instance, the average commute distance is 6 kilometres (4 miles) for small cities, and no more than 9.3 kilometres (5.8 miles) for super large-sized ones. Electric bikes have expanded the maximum distance people are willing to cover, with one-way distances of some 10–12 kilometres (7 miles) being no big deal, if you can do it safely. But even longer trips are possible by those means, and 'slow travel' is getting trendier. When walking or cycling don't meet your needs, or if the infrastructure is not yet there for you to do these safely, public transportation should always be the preferred choice. Depending on where you live, you may need to combine the two – like cycling to a suitable train stop.

➤ ***Fly less.***

Some countries, like Austria and France, have already banned short-haul flights (flights that can be replaced by train journeys of up to around 3 hours; other countries have different definitions). Several international companies are enforcing similar policies for their employees, although in most cases the decision to skip flying still rests with us. Overseas travel is one of the few areas where no other transportation means are readily available, and one where I've struggled to reduce my own footprint due to the international nature of my work. During the pandemic, however, I've been able to organise or participate in many fully virtual international workshops, conferences and other meetings. Additional benefits include increased accessibility for those with disabilities or economic and time constraints. I hope virtual or 'hybrid' events will continue to prevail across society.

➤ ***Share a low-emissions vehicle.***

If you really need a car, be aware that choosing the right one is not a trivial matter. The once-popular substitution of fossil fuels by 'renewable ethanol', despite being marketed as a green alternative, contributed to driving deforestation in the Amazon in order to grow sugar cane. As a newer and more promising alternative, electric cars have surged in popularity for being regarded as 'environmentally clean', which to a large extent is true, but they still remain resource-consuming to produce. Their batteries require some 20 minerals, including lithium, cobalt, nickel and rare earth metals obtained through mining – an activity that poses a direct threat to many species around the world, from the salt plains of Latin America to the rainforests of Guinea and the deep seas of Fiji. Mineral demand has skyrocketed in recent years, calling for urgent replacements, alongside full recycling. And don't forget that the electricity required to power electric cars also comes at an environmental cost, even if from renewable sources. If possible, join a car-sharing scheme rather than owning one outright: this is a great solution that is now available in most places. After all, the vast majority of cars are not used efficiently, as they stand idle most of the time.

SOFT POWER

In August 2018, 15-year-old Greta Thunberg started to skip school and sit outside the Swedish Parliament every Friday with a handwritten sign: 'School strike for climate'. No one could have guessed the incredible knock-on effect that her actions would have over the following years – raising awareness and triggering protests and demands from millions of people around the world. But you don't have to be Greta to influence others to make a positive impact. Never underestimate your potential!

➤ ***Influence your workplace.***
As an employee or employer, your potential to influence change can have a huge impact. Check that your workplace has a comprehensive and credible sustainability strategy,

and that it is being de facto implemented. If not, find ways to make it happen – talk to the right people, or do it yourself. Companies often have opportunities to make substantial changes to reduce their environmental footprint. Encourage your employer to join the race to **net zero** and take all possible steps to become climate positive on a rapid and science-based emissions reduction pathway. Employees have a very important voice, and employers should seek to upskill and empower their staff to understand the environmental impacts of their work, for example by providing training in carbon literacy and sustainable procurement. It's important to recognise that offsetting carbon emissions isn't a replacement for absolute reductions, and indiscriminately buying into certain carbon schemes could have a negative impact on biodiversity. Particular areas to consider are the purchasing of goods and services; the ability to repair, reuse and store equipment and resources; business travel policies; workplace food and drink; the construction of new buildings and refurbishment of old ones; enabling low-carbon commuting choices; and transparent reporting of company impacts.

➤ ***Ask the critical questions.***

'Where does this product come from?' 'How do you know the information on this label is true?' 'What does this ingredient mean?' The more pressure we put on the companies selling us services and goods that we find in shops and supermarkets, the more likely it will be that these companies will stop selling the products that damage the environment, or at least they will provide customers with transparent information to allow for informed decisions. For instance, far too many products have the cryptic ingredient 'vegetable oil' on their label, which is nearly always palm oil – a commodity that, as we've seen in Chapter 9, is the key driver of deforestation in places like Indonesia and Malaysia. Besides clearer reporting of a product's components, our requests will encourage companies to increase the transparency of supply chains, and to provide evidence-based assessments of their environmental and social impact. In Sweden, the mega-grocery chain COOP came up with a way to portray ten different metrics in a single symbol (Fig. 27), which is a good step forward despite the challenges of measuring them reliably.

Similar questions and requests can be made to the manifold organisations we interact with: our child's school, our choir, church, sports centre, entertainment park, concert hall, theatre and more. And don't be shy to take it a step further – writing to your local newspaper, phoning in to radio programmes, joining peaceful protests, setting up (or at least signing) petitions to reverse bad government decisions that may negatively affect biodiversity. The more we speak up, the easier and less awkward it'll feel to all of us, and the faster we'll change society.

➤ ***Vote strategically.***

Recent years have seen multiple examples of world leaders paying lip service to environmental issues. Even worse, some have actively denied climate change and

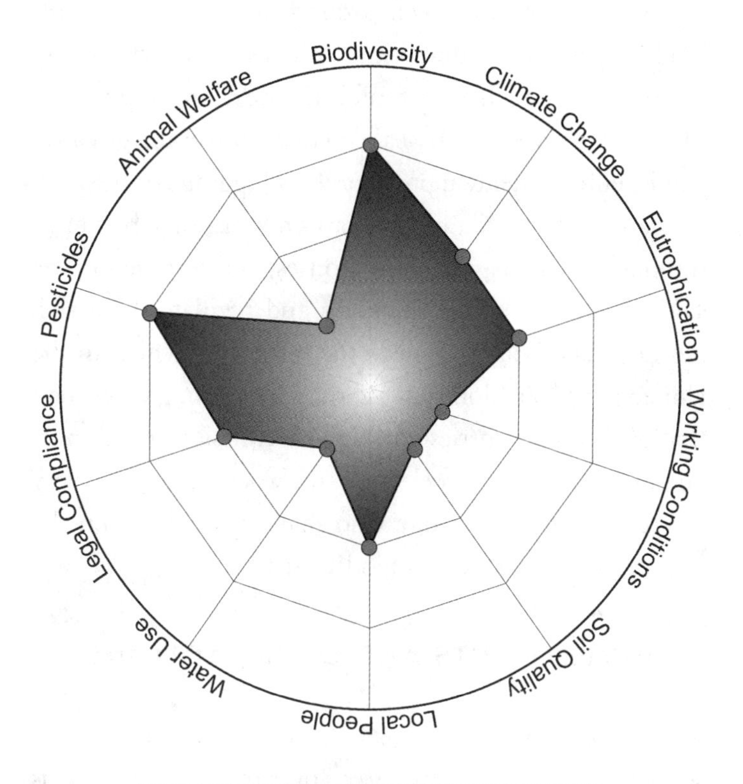

Figure 27. Disclosing the socio-environmental impact of a product. Since our consumption (particularly of food) is the most important driver of biodiversity loss, we need to know the impact of various products so that we can make informed decisions. This 'spider web' diagram shows the relative impact of an item on 10 different variables. For each variable, the further away a point is from the centre of the diagram, the larger the negative impact of the product. As consumers, we can demand transparency from supermarkets and other providers of products and services – you shouldn't need to be an environmental expert to consume sustainably!

made decisions that counter green pathways to economic development. Policymakers around the world, at various levels – presidents and prime ministers, governors and mayors – can all have a profound effect on biodiversity. They can strengthen or weaken environmental legislation and its enforcement, approve or deny logging concessions, increase or decrease taxes on carbon emissions, ban environmentally damaging commodities, and so much more. They decide on public spending and regulate the activities of public bodies, such as the military – which in the United Kingdom alone has a carbon footprint greater than that of the 60 countries with lowest emissions combined. So next time there's an election, make sure to choose very carefully to elect someone who shares your views on the importance of nature and biodiversity.

INVESTMENTS AND OTHER ACTIONS

➤ ***Donate money.***

We all pay taxes, but the way governments use them is unfortunately nowhere near sufficient to halt biodiversity loss and combat climate change around the world, particularly in low-income biodiverse countries. Luckily, there are many organisations doing incredible work on the ground, such as helping local communities to develop environmentally sustainable income sources, training and paying people to prevent illegal logging and hunting in protected areas, promoting environmental education among children, and much more. And this is really needed: in Africa, the number of forest elephants fell by more than 86 per cent during the past three decades, due to poaching and habitat degradation – problems affecting thousands of other species around the world. This means that biodiversity monitoring, often carried out by NGOs and other charities that rely on donations, can play a crucial role in safeguarding those species.

Most people in high-income countries should be able to give away 1 per cent of their monthly income to a cause they support, and even a 5 per cent contribution may go largely unnoticed. You could also donate to help compensate for the emissions of a particular action, such as a necessary flight. Donating to a voluntary offsetting initiative can support the restoration of nature, bringing tangible benefits to both climate and biodiversity (although this doesn't mean we can keep emitting carbon as usual). Finding the right initiative isn't always easy, but as a principle if you are supporting sustainable development, such as renewable power installations in developing countries, or the protection or restoration of a natural habitat, such as rainforest, mangroves, sea-grass beds or peatland, by an organisation or charity with the necessary expertise, you will be making a positive difference. Be wary of tree-planting initiatives if it is unclear whether they follow best practice – such as the ten golden rules I mentioned in the previous chapter, which consider what trees are being planted, and how they are being maintained.

Although many people like to decide on the exact use of their donations, I would recommend finding an organisation you trust, and letting it use the money where it is most needed, rather than on the item that attracts most public appeal. I have nothing against supporting cute and iconic animals such as pandas or tigers – they are phenomenal species and play key ecological roles – but there are many other species and ecosystems that require urgent protection and restoration. People have many causes they want to support, but environmental work really needs more cash, and is not receiving nearly as much as other areas. In the USA, for instance, 47 per cent of all donations go to social causes, 31 per cent to religious groups, and a mere 3 per cent to the environment.

* *Watch your savings.* Three out of four people have no idea where their pension money is being invested. Many countries and companies have adopted automatic enrolment, meaning that people's pensions go automatically into 'default' funds. Despite the convenience of this, it means that your hard-won savings could be used to support mining, oil and gas industries, or other environmentally damaging activities. Make sure to opt for ethical financial institutions with green portfolios, which are growing in popularity and can play a massive role in supporting transitions towards sustainability in the private sector.

➤ ***Record species sightings.***

A major prerequisite to protecting biodiversity is to understand where each species lives. Otherwise, building the next factory may destroy an entire population of a rare salamander, or kill a plant previously unknown to science. As our climate changes and habitats are modified, species reach new areas and disappear from others. Fortunately, it's never been so easy to help the global effort of mapping the world's biodiversity and how it's changing over time. If you own a smartphone, you can download an app called iNaturalist* and start logging your sightings straight away. You don't even need to know what species you are photographing, as the software will do a good job in matching your photo with millions of others, using artificial intelligence, and the user community can help verify the identification. This is a really fun activity to do with friends and family, be it around your house or during hikes in nature and travels; you'll soon learn to recognise lots of species with minimal effort. Millions of species observations have already been made by a large community of people, but many more are needed: all our sightings matter.

➤ ***Stay curious.***

If you've read this far, you now know more about biodiversity, its values, threats and solutions than the vast majority of people. But don't stop here. If there are particular aspects that appeal to you – perhaps you'd like to learn more about a certain group of species, or how to support biodiversity

---

* There are other alternatives too; one important thing is that the app doesn't provide users with precise coordinates of rare or threatened species with commercial value, which could be misused by poachers – a growing problem in South Africa, for instance.

action in your community, or become a biodiversity scientist yourself – go ahead! The world is in desperate need of advocates for nature, and change starts with you: you can really make a difference.

EPILOGUE: LOOKING AHEAD

As a child, I grew up admiring the endless beauty of nature: from the unthinkable depths and countless stars of the universe – which my father and I explored through the lenses of our small telescope while conversing long into the warm nights in Brazil – to the myriad life forms that surrounded me in the world's most biodiverse country. How fortunate I was, and we all are, to share this incredible planet! And despite much speculation, we actually have no idea if there might be anything else like this 'out there'. If there is, it would be so far away that it would take many millions of years to get there, and it is very unlikely that we would feel at home. So let us not give up on Earth, whose biological complexity is a universe of its own, most of which remains to be unveiled, but is disappearing faster than ever before in human history.

One of the most common questions I get is whether I'm optimistic, and if I have hope. As a scientist, my views are shaped by evidence: how things are today, how they have changed over time, and what mathematical models can predict. The truth is, things are looking really bad, and the prospects are dire. We are *not* anywhere close to halting, let alone reversing, biodiversity loss. In a study led by my former PhD student Tobias Andermann, we estimated that the activity of humans has already led to a 1,700-fold increase in extinction rates of mammals, when compared to natural levels. If current trends continue, this will increase to 30,000 times by the end of this century.

Optimism and hope, therefore, are quite irrelevant here. What really matters is action. In 1972, Swedish prime minister Olof Palme invited the world's leaders to the first meeting of the United Nations on the environment, where he urged them to join forces in urgently tackling the already ongoing environmental destruction. Since then, through treaties and conventions, nations have repeatedly agreed to ambitious targets in pursuit of that goal, but have nearly consistently failed in delivering them. In 2010, 194 countries and territories committed to halt biodiversity loss through 20 concrete targets with a deadline of 2020. In reality, at the end of that period *none* of the goals were fully met.

When will all of us realise that we are shooting ourselves in the foot? Between 2001 and 2020, 411 million hectares of forest – an area twice the size of Mexico – were lost globally due to deforestation, mainly driven by the expansion of agriculture. At the same time, over 90 per cent of the world's poorest people depend on forests for their livelihoods, and so their disappearance is jeopardising their futures.

This cannot continue. The time has come to make the protection of biodiversity and restoration of degraded ecosystems the primary focus across all segments of society.

It must be acted on by all of us, in our homes and families, in our local communities, our national governments, and on the global stage. This is so critical to our future that the United Nations has placed biodiversity at the core of its vision for the future of our planet, under the framework of the Sustainable Development Goals (Goal 14: Life Below Water and Goal 15: Life on Land).

One might question this call for massive investments – both in time and money – in our natural world, when so many other social and economic challenges abound. But focusing the world's attention and resources on protecting and restoring nature will not only help to stop the catastrophic loss of biodiversity we're witnessing today; it will also directly benefit all of us. Safeguarding biodiversity will contribute directly to supporting sustainable livelihoods, so that we can all live a worthy and healthy life; it will improve global food security, so that we are less likely to experience widespread famine and drought, which are main drivers of human displacement, social conflict and war; it will safeguard people's access to key – and sometimes rare – medicinal plants, to help treat diseases and save lives; it will help protect watersheds, to maintain and regulate our natural habitats and provide clean water for people and agriculture; and it will increase our resilience under climate change, which will continue to test us and our survival for decades and centuries to come.

This is not an exhaustive list – the benefits of protecting and restoring our hidden universe are, like the stars, too numerous to mention one by one. But if we are to continue living and surviving together as a human species, we must take stock of what the natural world gives us, and needs from us, before it all disappears and it becomes too late.

To be honest, I *do* hold both huge optimism and hope – in equal doses with anxiety. It is those three feelings that get me out of bed each morning. They have also made me choose my profession, and they have spurred me on to write this book. My *hope*, however, is not that 'everything will be fine, no matter what'; rather, it is that enough of us, including our leaders, will realise that there is no other option than to completely change our lifestyle, and get our priorities right. My *optimism* is that this societal transformation – underpinned by new technological advances and nature-based solutions, and the actions I listed in this final part of the book – will simultaneously benefit our wellbeing and our planet. My *anxiety* is that this transformation will take far too long, and in that process this world will have lost far too many of its natural ecosystems and species – including perhaps our own.

The numerous species we have already obliterated, in our quest to colonise and multiply in every corner of this planet, are gone forever. But as long as a million species today are only threatened and not completely gone, opportunities remain. Yes, there *is* a big risk, but it *is* possible to reverse this trend. In fact, it can go incredibly fast – if we do it right, and start immediately. The decisions we make today will influence the fate of biodiversity, and our planet, for millions of years.

And despite all the stupid mistakes we have made in the past, I believe we all want to choose a different future. A future where we make peace with nature. A future where we do not take more than we really need, and restore what we have taken. A future where the fantastic forests I played in as a child in Brazil, collecting seeds and insects in shoeboxes, remain for all future generations to come, to continue playing in and marvelling at. A future where we finally realise that we are also an animal – a part of nature, not separate from it.

We are a species that unfortunately has managed to develop the ability to destroy its own universe: our natural home. But, fortunately, we are a species that has also the ability to put everything right again – if we only want to. The solution, I'm convinced, is in all of us.

ACKNOWLEDGEMENTS

As a scientist, it's been an adventure to write a popular science book – a bit like venturing into the deep rainforests of the Amazon or hiking in the eastern African mountains, in the search for new or poorly known species. Luckily, I received a great deal of encouragement and support to make this happen.

I am indebted to my editors Albert DePetrillo and Hana Teraie-Wood for believing in the book and helping co-develop it; Rhian Smith, Joseph Calamia, Claes Bernes, Michael Bright, Heather McLeod, Josephine Maxwell for excellent feedback and editorial support; Gina Fullerlove, Ciara O'Sullivan, Michael McCarthy and Allison Perrigo for their encouragement right from the beginning; Richard Deverell and Sandra Botterell for getting fully behind the idea; Lizzie Harper and Meghan Spetch for their great illustrations, with additional help from Harith Farooq and Stephen Smith; Martyn Ainsworth, Mónica Arakaki, Elinor Breman, Bethanie Carney Almroth, William Baker, Nataly Canales, Paul Cannon, Mark Chase, Carly Cowell, Aaron Davis, Victor Deklerck, Sam Dupont, Johan Eklöf, Christer Erséus, Oscar Pérez Escobar, Kate Evans, Søren Faurby, Harith Farooq, Peter Gasson, Kate Hardwick, Ulf

Jondelius, Gareth Jones, Kirsten Knudsen, Matthias Obst, Carla Maldonado, Mark Nesbitt, Tuula Niskanen, Catalina Pimiento, Rachel Purdon, Hélène Ralimanana, Ferran Sayol, Per Sundberg, Maria Vorontsova and Kim Walker for fact-checking and expert advice on specific topics of this book; my great colleagues and friends at Kew, in Gothenburg, in Brazil and several other countries, who are just too many to mention here but who have generously shared their knowledge over the years and helped shape my ideas on this topic.

Finally, all my love and thanks go to my wife Anna and our children Maria, Clara and Gabriel Antonelli for our numerous dinner conversations on biodiversity; for helping to co-develop and put in practice many pieces of advice on solutions to promote a more environmentally sustainable lifestyle at home and beyond; and for our joint decision to allocate our multi-year family savings to protect and restore rainforest in Brazil, a project to which I will be sending all my proceeds from this book.

GLOSSARY

Archaea: an ecologically important but poorly known branch on the *Tree of Life*, whose tiny species share some traits with bacteria (such as consisting of a single cell and lacking a nucleus) but whose DNA is more similar to that of organisms containing a nucleus (such as plants, fungi and animals).

Biodiversity: The variety of all life on Earth. This word is a contraction of 'biological diversity' and comprises at least five components: species diversity, genetic diversity, evolutionary (or phylogenetic) diversity, functional diversity and ecosystem diversity.

Biogeography: The science that aims to document and understand how biodiversity is distributed across the world and how it changes over time.

Biomimetics: The pursuit of solutions inspired by or copied from nature, to tackle various engineering problems and other societal needs. Also called biomimicry.

Biota: The full set of species belonging to a particular region, such as an island or ecosystem.

Circular production: The manufacturing of products and services that seek to reuse and recycle materials and energy, in order to achieve long-term environmental sustainability and avoid the need of further extraction of natural resources.

Climatic tolerance: The set of climatic conditions – such as ranges of temperature, rainfall and seasonality – that a species or individual tolerates or finds optimal for survival.

Colonisation: The arrival and successful establishment of a species into a new region or habitat it did not previously occur in, such as a fish colonising a new lake or a bird arriving to a new continent.

Convention on Biological Diversity: The set of international agreements signed by the vast majority of countries with the aim of protecting and sustainably using the world's biodiversity, with a fair and equitable sharing of benefits derived from it.

Convergent evolution: The natural phenomenon whereby some distantly related species (or groups of organisms) come to resemble one another as a result of evolution under similar environmental pressures. Examples include certain American cacti and African euphorbias, which have adapted to dry environments in similar ways, and dolphins and tuna, whose streamlined bodies are optimised for fast swimming over long periods.

Cryptic species: A species which appears to be easily recognisable in form but is in fact composed of multiple distinct evolutionary entities that may deserve separate species status, as revealed by detailed genetic or ecological analyses.

Ecocide: A proposed international crime to recognise severe and widespread damage to the environment, caused by unlawful or wanton acts by individuals, companies or governments.

Ecosystem: An assemblage of species interacting with the physical components of a particular environment. Examples are tropical rainforests, savannahs and coral reefs.

Ecosystem engineers: Species that play a key role in shaping an *ecosystem*, such as beavers building dams and woodpeckers whose search for food creates nests for other species of birds and small mammals.

Ecosystem services: The vast range of benefits that nature provides us with, which are intrinsically linked to healthy, biodiverse ecosystems. These services are usually grouped into provisioning (such as food, medicines, fibres, building material), cultural (recreation, ecotourism) and regulating (pollination, water purification, climate regulation, flood control, carbon sequestration). See also *nature's contributions to people.*

Evolutionary / phylogenetic diversity: The sum of evolutionary history captured by a set of species, often measured as the total amount of time that has elapsed since they shared a common ancestor, or the genetic variation they have accumulated among themselves.

Extinction debt: The number of species in a region that are bound to go extinct as the result of previous environmental degradation. This number can be predicted from species-area relationships and other biological factors, such as the inherent genetic diversity of species and their requirements for space, food sources and partners for reproduction.

Form: See **Morphology**.

Fossil record: The sequence of extinct organisms preserved in rock sediments, which can inform on how species, ecosystems and life forms have changed over time and in different regions.

Function: The role that a species or group of species plays in an ecosystem, often associated with a particular morphology (called a trait) that influences an ecosystem, by interacting with other species and the environment. For example, many fungi function as decomposers in their ecosystems; cats and their relatives are carnivores (meat eaters); and antelopes and grasshoppers are both herbivores (plant eaters).

Functional diversity: The total variety of ecological functions in a particular system, such as an island or a lake.

Genetic diversity: The total variety in genetic material – including DNA sequences, genes and alleles (variants of genes) – among individuals belonging to the same population or species.

Genome. The entire set of chromosomes, and the genetic information they contain, within an organism, consisting of DNA molecules (or RNA in viruses). This includes both genes that are used to produce proteins (called 'coding' genes) and those that do not code and whose functions, if any, remain disputed.

The **Great Acceleration:** The fast and ubiquitous surge in multiple metrics related to human activity since the 1950s, such as population growth, deforestation rates, atmospheric greenhouse gases and agricultural land used. Some of these metrics have declined in recent years.

Herbarium specimen: The dried, pressed sample of a plant (in whole or in part) mounted on a large sheet alongside a label containing the species name, the precise location and date the specimen was collected, the collector's name and any other pertinent information. Collections of herbarium specimens are kept in herbaria and are widely used as scientific reference material.

Invasive species: A species introduced from another region (also called non-native, or alien) that multiplies and spreads in a way that harms the invaded ecosystem.

Keystone species: A species with a disproportionate influence on other species in its ecosystem, such as lions in savannahs. Removal of a keystone species will lead to far-reaching effects on the diversity and abundance of other species and can fundamentally change how an ecosystem functions.

Latitudinal diversity gradient: The fact that most organism groups – such as birds, plants and insects – have the highest species richness near the Equator, with their diversity gradually decreasing towards higher altitudes of the northern and southern hemispheres.

Life forms: Another term for organism groups, such as species.

Microbiome: All the microorganisms (such as bacteria, viruses and fungi) that co-exist in an organism or organ, such as the human gut.

Microplastics: Tiny bits of plastic (smaller than 5 millimetres) resulting from the breakdown of plastic products and waste, such as fleece jackets and plastic bags. Fragments smaller than 1000 nanometres (0.001 millimetres) are called nanoplastics.

Morphology: The form, shape or structure of a species. Most species are morphologically distinct from one another, although some may be *cryptic.* Also called *form.*

Natural capital: The natural assets of a region or ecosystem, including all its living organisms, soil, water, air and minerals.

Nature's contributions to people. Another term for *ecosystem services,* which more clearly includes the wide non-material contributions provided by ecosystems and their biodiversity to the quality of life for people, such as spiritual, cultural, recreational and other values.

Net zero: A term referring to neutral impact, often used for describing the goal of avoiding increases in carbon emissions over time as a means to mitigate global climate warming through a reduction in emissions and removal of previously released carbon dioxide.

Ocean acidification: The increase in sea water acidity as a result of the higher concentrations of carbon dioxide in the atmosphere, resulting in negative impacts on marine species.

Organism groups: A term used to refer to related organisms – usually applied to levels of classification above species. For example, frogs (including all constituent species) form an organism group, and vertebrates (animals with backbones) are a broader organism group that includes frogs.

Phenology: The study of when natural phenomena occur in nature, such as the flowering and fruiting timing of trees and the seasonal migration of certain fish.

Phylogeny: An evolutionary tree showing how species are related – also called a Tree of Life.

Produced capital. Goods or structures produced by humans, such as roads, buildings, machines and other forms of infrastructure.

Range: The geographic area in which a species occurs. 'Native' range refers to the natural distribution of a species, while 'introduced' or 'naturalised' range refers to a region where a species is found as a result of human-mediated or accidental introduction.

Redundancy: The fact that many species in a healthy ecosystem tend to exhibit similar ecological functions, such as insect-eating birds in a savannah or excavating worms in a sandy beach.

Speciation: The formation of new species.

Species: The fundamental and most widely used unit of biodiversity, species are commonly defined as comprising all individuals that can exchange genes with each other through sexual reproduction. However, many alternative species concepts exist, and no universal criteria can be applied to every organism group.

Species richness: The number of species in a region. Also called alpha diversity, taxonomic diversity or species diversity.

Species-area relationship: The well-documented statistical relationship between the size of an area and the number of species it will naturally contain, all else being the same.

Subspecies: A subcategory of species, used to define a set of individuals or populations that are likely to evolve into a different species over time, but where members of different subspecies are still able to reproduce with one another successfully.

Taxonomy: The scientific discipline of naming, describing and classifying organisms such as species.

Technosphere: The human-built part of the world, comprising buildings, roads, machines, railways, oil platforms, artificial satellites and other artefacts.

Tree of Life: The graphical representation of how species are related to one another through common descent, as inferred from the study or their genetic or morphological differences. Another term for phylogeny.

FURTHER READING

I have made my best attempt to provide accurate information throughout this book, while avoiding jargon and the complexity always associated with the natural world and scientific research. Below I provide some general and specific suggestions for further reading, including the sources of information for key statements in the various chapters. Please note that not all of them have their full texts freely accessible, particularly scientific articles; but you can nearly always read their summary (abstract) for free, or request a copy from a university library or the authors directly.

➤ ***Kew's mission:***
Royal Botanic Gardens, Kew, *Our manifesto for change 2021-2030* (2021)
Available at: www.kew.org

➤ ***My own research:***
Antonelli Lab, *antonelli-lab.net*
Available at: http://antonelli-lab.net and http://tiny.cc/antonelli

➤ ***Explore the topics covered further at Kew's Library and Archives:***
Royal Botanic Gardens, Kew, 'Library and Archives.'
kew.org

Available at: www.kew.org/kew-gardens/whats-in-the-gardens/library-art-and-archives

➤ ***Access Kew's collections catalogues:***
Royal Botanic Gardens, Kew, 'Collections Catalogues.' *kew.org*
Available at: www.kew.org/science/collections-and-resources/data-and-digital/collections-catalogues

SETTING THE SCENE

➤ ***Astronomical discoveries, historical and more recent research:***
Alfred, R., 'Dec. 30, 1924: Hubble Reveals We Are Not Alone.' *Wired* (30 December 2009)
Available at: www.wired.com/2009/12/1230hubble-first-galaxy-outside-milky-way/

Johnson, G., 'Miss Leavitt's Stars: The Untold Story of the Woman Who Discovered How to Measure the Universe'. (W. W. Norton Company, 2015)

NASA, Goddard Space Flight Center, 'Biography of Edwin Powell Hubble (1889 – 1953).' *Nasa.gov*
Available at: asd.gsfc.nasa.gov/archive/hubble/overview/hubble_bio.html

NASA, 'Dark Energy, Dark Matter'. *Science.nasa.gov* (October, 2021)
Available at: https://science.nasa.gov/astrophysics/focus-areas/what-is-dark-energy

Siegel, Ethan, Starts With A Bang, 'How Much Of The Unobservable Universe Will We Someday Be Able To See?'

Forbes (5 March 2019)
Available at: https://www.forbes.com/sites/startswithabang/2019/03/05/how-much-of-the-unobservable-universe-will-we-someday-be-able-to-see/

➤ ***Early human exploration of biodiversity:***

Ben-Dor, M. *et al.*, 'Man the Fat Hunter: The Demise of Homo erectus and the Emergence of a New Hominin Lineage in the Middle Pleistocene (ca. 400 kyr) Levant.' *PLOS ONE* 6 (2011): e28689
doi: 10.1371/journal.pone.0028689

Brumm, A. *et al.*, 'Age and context of the oldest known hominin fossils from Flores.' *Nature* 534 (2016): 249–253
doi: 10.1038/nature17663

Diamond, J. M., *Guns, Germs and Steel: The Fates of Human Societies* (Jonathan Cape, 1997)

Pan, S.-Y. *et al.*, 'Historical perspective of traditional indigenous medical practices: the current renaissance and conservation of herbal resources.' *Evidence-Based Complementary and Alternative Medicine* (2014): 525340
doi: 10.1155/2014/525340

➤ ***Linnaeus's work:***

Blunt, W., *Linnaeus: The Complete Naturalist* (Princeton University Press, 2002)

'Carolus Linnaeus.' *Britannica* (2021)
Available at: https://www.britannica.com/biography/Carolus-Linnaeus

➤ ***Tree of life:***
Baker, W. J. *et al.*, 'A Comprehensive Phylogenomic Platform for Exploring the Angiosperm Tree of Life.' preprint. *Evolutionary Biology* (2021)
doi: 10.1101/2021.02.22.431589

Hinchliff, C. E. *et al.*, 'Synthesis of phylogeny and taxonomy into a comprehensive tree of life.' *Proceedings of the National Academy of Sciences* 112 (2015): 12764–12769
doi: 10.1073/pnas.1423041112

➤ ***Marine inventories in Sweden:***
Obst, M. *et al.*, 'Marine long-term biodiversity assessment suggests loss of rare species in the Skagerrak and Kattegat region.' *Marine Biodiversity* 48 (2018): 2165–2176
doi: 10.1007/s12526-017-0749-5

Willems, W. *et al.*, 'Meiofauna of the Koster-area, results from a workshop at the Sven Lovén Centre for Marine Sciences (Tjärnö, Sweden).' *Meiofauna Marina* 17 (2009): 1–34

➤ ***Estimates of species diversity and discovery:***
Costello, M. J. *et al.*, 'Can we name Earth's species before they go extinct?'. *Science 339(6118):413-6.* doi: 10.1126/science.1230318

Locey, K. J. and Lennon, J. T., 'Scaling laws predict global microbial diversity.' *Proceedings of the National Academy of Sciences* 113 (2016): 5970–5975
doi: 10.1073/pnas.1521291113

Mora, C. *et al.*, 'How Many Species Are There on Earth and in the Ocean?' *PLOS Biology* 9 (2011): e1001127
doi: 10.1371/journal.pbio.1001127

Wu, B. *et al.*, 'Current insights into fungal species diversity and perspective on naming the environmental DNA sequences of fungi.' *Mycology* 10 (2019): 127–140
doi: 10.1080/21501203.2019.1614106

➤ ***The human microbiome:***

Gilbert, J. *et al.*, 'Current understanding of the human microbiome.' *Nature medicine* 24 (2018): 392–400
doi: 10.1038/nm.4517

Huttenhower, C. *et al.*, 'Structure, function and diversity of the healthy human microbiome.' *Nature* 486 (2012): 207–214
doi: 10.1038/nature11234

'NIH Integrative Human Microbiome Project .' (2021)
Available at: https://hmpdacc.org/ihmp/

Yatsunenko, T. *et al.*, 'Human gut microbiome viewed across age and geography.' *Nature*, 486 (2012): 222–227
doi: 10.1038/nature11053

➤ ***Insect diversity in trees:***

Erwin, T. L. and Scott, J. C., 'Seasonal and Size Patterns, Trophic Structure, and Richness of Coleoptera in the Tropical Arboreal Ecosystem: The Fauna of the Tree Luehea seemannii Triana and Planch in the Canal Zone of Panama.' *The Coleopterists Bulletin* 34 (1980): 305–322

➤ ***The multi-dimensional aspects of biodiversity:***

Swenson, N. G., 'The role of evolutionary processes in producing biodiversity patterns, and the interrelationships between taxonomic, functional and phylogenetic biodiversity.' *American Journal of Botany* 98 (2011): 472–480
doi: 10.3732/ajb.1000289

➤ ***Indigenous biodiversity knowledge:***
Berlin, B., *Ethnobiological Classification: Principles of Categorization of Plants and Animals in Traditional Societies* (Princeton University Press, 1992)

Gillman, L. N. and Wright, S. D., 'Restoring indigenous names in taxonomy.' *Communications Biology* 3 (2020): 1–3
doi: 10.1038/s42003-020-01344-y

CHAPTER 1

➤ ***UK bats:***
Barlow, K. E. and Jones, G., 'Pipistrellus nathusii (Chiroptera: Vespertilionidae) in Britain in the mating season.' *Journal of Zoology* 240 (1996): 767–773
doi: 10.1111/j.1469-7998.1996.tb05321.x

Bat Conservation Trust:
www.bats.org.uk

Jones, G. and Van Parijs, S. M., 'Bimodal Echolocation in Pipistrelle Bats: Are Cryptic Species Present?' *Proceedings: Biological Sciences* 251 (1993): 119–125

➤ ***Orchid pollination:***
Antonelli, A. *et al.* 'Pollination of the Lady's slipper orchid (Cypripedium calceolus) in Scandinavia – taxonomic and conservational aspects'. *Nordic Journal of Botany* 27(4): 266-273 (2019).

Knapp, S., *Extraordinary Orchids*. Chicago University Press (2021)

➤ ***Scientific discoveries of fungi:***

Cheek, M. *et al.*, 'New scientific discoveries: Plants and fungi.' *PLANTS, PEOPLE, PLANET* 2 (2020): 371–388
doi: 10.1002/ppp3.10148

Douglas, B., 'The Lost and Found Fungi project.' *Kew Read & Watch* (1 February 2016) Available at:
www.kew.org/read-and-watch/lost-and-found-fungi

➤ ***World's largest organism:***

Anderson, J. B. *et al.*, 'Clonal evolution and genome stability in a 2500-year-old fungal individual.' *Proceedings of the Royal Society B: Biological Sciences* 285 (2018): 20182233
doi: 10.1098/rspb.2018.2233

➤ ***Mammoth DNA:***

van der Valk, T. *et al.*, 'Million-year-old DNA sheds light on the genomic history of mammoths.' *Nature* 591 (2021): 265–269
doi: 10.1038/s41586-021-03224-9

➤ ***Gene flow among species:***

Jónsson, H. *et al.*, 'Speciation with gene flow in equids despite extensive chromosomal plasticity.' *Proceedings of the National Academy of Sciences* 111 (2014): 18655–18660

Lexer, C. *et al.*, 'Gene flow and diversification in a species complex of Alcantarea inselberg bromeliads.' *Botanical Journal of the Linnean Society* 181 (2016): 505–520
doi: 10.1111/boj.12372

➤ ***Neanderthal and human interbreeding:***

Green, R. E. *et al.*, 'A Draft Sequence of the Neandertal Genome.' *Science* 328 (2010): 710–722
doi: 10.1126/science.1188021

➤ ***Global species observations:***
GBIF: The Global Biodiversity Information Facility, '*What is GBIF?*' (2021)
Available at: www.gbif.org/what-is-gbif

➤ ***The Theory of Island Biogeography:***
Drakare, S., Lennon, J. J. and Hillebrand, H., 'The imprint of the geographical, evolutionary and ecological context on species–area relationships.' *Ecology Letters* 9 (2006): 215–227
doi: 10.1111/j.1461-0248.2005.00848.x

MacArthur, R., Wilson, E.O., *The Theory of Island Biogeography.* Princeton University Press (1967)

➤ ***Tree diversity in the Amazonia:***
ter Steege, H. *et al.*, 'Hyperdominance in the Amazonian Tree Flora.' *Science* 342 (2013): 1243092
doi: 10.1126/science.1243092

ter Steege, H. *et al.*, 'Towards a dynamic list of Amazonian tree species.' *Scientific Reports* 9 (2019): 3501
doi: 10.1038/s41598-019-40101-y

Valencia, R., Balslev, H. and Paz Y Miño C G., 'High tree alpha-diversity in Amazonian Ecuador.' *Biodiversity & Conservation* 3 (1994): 21–28
doi: 10.1007/BF00115330

➤ ***The rarity of species:***
Enquist, B. J. *et al.*, 'The commonness of rarity: Global and future distribution of rarity across land plants.' *Science Advances* 5 (2019): eaaz0414
doi: 10.1126/sciadv.aaz0414

Zizka, A. *et al.*, 'Finding needles in the haystack: where to look for rare species in the American tropics.' *Ecography*, 41 (2018): 321–330
doi: 10.1111/ecog.02192

CHAPTER 2

➤ ***Coffee research:***

Borrell, J. S. *et al.*, 'The climatic challenge: Which plants will people use in the next century?' *Environmental and Experimental Botany* 170 (2020): 103872
doi: 10.1016/j.envexpbot.2019.103872

Moat, J. *et al.*, 'Resilience potential of the Ethiopian coffee sector under climate change.' *Nature Plants* 3 (2017): 1–14
doi: 10.1038/nplants.2017.81

Davis, A.P. *et al.*, 'Arabica-like flavour in a heat-tolerant wild coffee species'. *Nature Plants* 7 (2021): 413–418.
doi: 10.1038/s41477-021-00891-4

➤ ***Ash dieback:***

Hill, L. *et al.*, 'The £15 billion cost of ash dieback in Britain.' *Current Biology* 29 (2019): R315–R316
doi: 10.1016/j.cub.2019.03.033

Stocks, J. J. *et al.*, 'Genomic basis of European ash tree resistance to ash dieback fungus.' *Nature Ecology & Evolution* 3 (2019): 1686–1696
doi: 10.1038/s41559-019-1036-6

➤ ***Galapagos finches:***

Grant, P. R. and Grant, B. R., 'Unpredictable Evolution in a 30-Year Study of Darwin's Finches.' *Science* 296 (2002): 707–711
doi: 10.1126/science.1070315

Ahmed, F. 'Profile of Peter R. Grant'. *PNAS* 107 (13) (2010): 5703–5705.
https://doi.org/10.1073/pnas.1001348107

➤ ***Genetic diversity in humans and fruit flies:***

Condon, M. A. *et al.*, 'Hidden Neotropical Diversity: Greater Than the Sum of Its Parts.' *Science* 320 (2008): 928–931
doi: 10.1126/science.1155832

National Institutes of Health (US) and Biological Sciences Curriculum Study, *Understanding Human Genetic Variation* (National Institutes of Health (US), 2007).

➤ ***Average longevity of mammals:***

Hagen, O. *et al.*, 'Estimating Age-Dependent Extinction: Contrasting Evidence from Fossils and Phylogenies'. *Systematic Biology* 67(3): 458–473
doi: 10.1093/sysbio/syx082

➤ ***Ancient DNA of date palm:***

Pérez-Escobar, O. A. *et al.*, 'Archaeogenomics of a ~2,100-year-old Egyptian leaf provides a new timestamp on date palm domestication.' preprint. *bioRxiv* (2020)
doi: 10.1101/2020.11.26.400408

➤ ***Kew Millennium Seed Bank Partnership:***

'Kew Millennium Seed Bank.', *kew.org*
Available at: www.kew.org/wakehurst/whats-at-wakehurst/millennium-seed-bank

'Celebrating 20 years of the Millennium Seed Bank and Millennium Seed Bank Partnership.' *Samara* 36 (2020): 1-20

CHAPTER 3

➤ ***Extinction of the thylacine:***
Boyce, J., 'Canine Revolution: The Social and Environmental Impact of the Introduction of the Dog to Tasmania.' *Environmental History* 11 (2006): 102–129
doi: 10.1093/envhis/11.1.102

Brass, K., 'The $55,000 search to find a Tasmanian tiger.' *Australian Women's Weekly* (24 September 1980): 40-41

'Thylacine.' *Britannica* (2021)
Available at: https://www.britannica.com/animal/thylacine

➤ ***Time estimates among species:***
Kumar, S. *et al.*, 'TimeTree: A Resource for Timelines, Timetrees, and Divergence Times.' *Molecular Biology and Evolution* 34 (2017): 1812–1819
doi: 10.1093/molbev/msx116

➤ ***Different ways of calculating phylogenetic (evolutionary) diversity:***
Tucker, C. M. *et al.*, 'A guide to phylogenetic metrics for conservation, community ecology and macroecology.' *Biological Reviews* 92 (2017): 698–715
doi: 10.1111/brv.12252

CHAPTER 4

➤ ***Experimental research on mountains:***
The GLORIA Network:
www.gloria.ac.at/network/general

Swiss Federal Institute for Forest, Snow and Landscape Research WSL, 'International Tundra Experiment ITEX.' *wsl.ch*
Available at: www.wsl.ch/en/projects/tundra-experiment

➤ ***Functional diversity and traits:***

Lefcheck, J., 'What is functional diversity, and why do we care?' *sample (ECOLOGY)* (20 October 2014).
Available at: jonlefcheck.net/2014/10/20/what-is-functional-diversity-and-why-do-we-care-2/

Shi, Y. *et al.*, 'Tree species classification using plant functional traits from LiDAR and hyperspectral data.' *International Journal of Applied Earth Observation and Geoinformation* 73 (2018): 207–219
doi: 10.1016/j.jag.2018.06.018

Stuart-Smith, R. D. *et al.*, 'Integrating abundance and functional traits reveals new global hotspots of fish diversity.' *Nature* 501 (2013): 539–542
doi: 10.1038/nature12529

CHAPTER 5

➤ ***Humboldt's travels and legacy:***

'Alexander von Humboldt Anniversary collection.' *nature ecology & evolution* (30 August 2019)
Available at: www.nature.com/collections/ceaeaabjia/

'Humboldt's legacy.' *Nature Ecology & Evolution* 3 (2019): 1265–1266
doi: 10.1038/s41559-019-0980-5

Journal of Biogeography 46(8) (2019): i-iv, 1625-1900

Rooks, T., 'How Alexander von Humboldt put South America on the map.' *Deutsche Welle (DW)* (12 July 2019)
Available at: p.dw.com/p/39v70

Wulf, A., *The Invention of Nature: The Adventures of Alexander von Humboldt, the Lost Hero of Science* (Hodder & Stoughton: 2015)

➤ ***Ecosystems and their borders:***

Antonelli, A., 'Biogeography: Drivers of bioregionalization.' *Nature Ecology & Evolution* 1 (2017): 0114
doi: 10.1038/s41559-017-0114

Arakaki, M. *et al.*, 'Contemporaneous and recent radiations of the world's major succulent plant lineages'. *Proceedings of the National Academy of Sciences* 108 (20): 8379–8384 (2011).
doi: 10.1073/pnas.1100628108

'Köppen climate classification.' *Britannica* (2021)
Available at: https://www.britannica.com/science/Koppen-climate-classification

➤ ***Shifts in large ecosystem regimes:***

Cooper, G. S., Willcock, S. and Dearing, J. A., 'Regime shifts occur disproportionately faster in larger ecosystems.' *Nature Communications* 11 (2020): 1175
doi: 10.1038/s41467-020-15029-x

➤ ***Death of the Aral Sea:***

Synott, M., 'Sins of the Aral Sea.' *National Geographic* (1 June 2015)
Available at: www.nationalgeographic.com/magazine/article/vanishing-aral-sea-kazakhstan-uzbekistan

CHAPTER 6

➤ ***Work on the coffee family, including quinine, by colleagues and myself (selection):***

Andersson, L. and Antonelli, A., 'Phylogeny of the tribe Cinchoneae (Rubiaceae), its position in Cinchonoideae, and description of a new genus, Ciliosemina.' *TAXON* 54 (2005): 17–28
doi: 10.2307/25065412

Antonelli, A. *et al.*, 'Tracing the impact of the Andean uplift on Neotropical plant evolution.' *Proceedings of the National Academy of Sciences* 106 (2009): 9749–9754
doi: 10.1073/pnas.0811421106

Traverso, V., 'The tree that changed the world map'. *BBC Travel.*
Available at: https://www.bbc.com/travel/article/20200527-the-tree-that-changed-the-world-map

Walker, K., Nesbitt, M., *Just the Tonic: A natural history of tonic water.* (Royal Botanic Gardens, Kew, 2019)

➤ ***Horseshoe crabs:***

Arnold, C., 'Horseshoe crab blood is key to making a COVID-19 vaccine—but the ecosystem may suffer.' *National Geographic* (2 July 2020)
Available at: www.nationalgeographic.com/animals/article/covid-vaccine-needs-horseshoe-crab-blood

➤ ***Manifold uses of plants and fungi:***

Antonelli, A. *et al.*, *State of the World's Plants and Fungi 2020* (Royal Botanic Gardens, Kew, 2020)
doi: 10.34885/172

Royal Botanic Gardens, Kew (ed.), 'Special Issue: Protecting and sustainably using the world's plants and fungi.' *PLANTS, PEOPLE, PLANET* 2 (2020): 367-579
Available at: nph.onlinelibrary.wiley.com/toc/25722611/2020/2/5

➤ ***Plants as sources of micronutrients:***

'11 Plant-Based Foods Packed With Zinc.' *EcoWatch* (7 April 2016)
Available at: www.ecowatch.com/11-plant-based-foods-packed-with-zinc-1891079003.html

Thomas, L., 'Sources of Selenium.' *News-Medical.net* (14 April 2021)
Available at: www.news-medical.net/health/Sources-of-Selenium.aspx

Ware, M., 'Selenium: What it does and how much you need.' *Medical News Today* (19 May 2021)
Available at: www.medicalnewstoday.com/articles/287842

➤ ***Banana blight:***

Dita, M. *et al.*, 'Fusarium Wilt of Banana: Current Knowledge on Epidemiology and Research Needs Toward Sustainable Disease Management.' *Frontiers in Plant Science* 9 (2018): 1468
doi: 10.3389/fpls.2018.01468

Espiner, T., 'Do we need to worry about banana blight?' *BBC News* (15 August 2019)
Available at: www.bbc.com/news/business-49331286

FAO (Food and Agriculture Organization), 'Banana facts and figures.' *fao.org*
Available at: www.fao.org/economic/est/est-commodities/bananas/bananafacts#.YOhfUehKg2w

➤ ***Biodiversity as an asset:***
Dasgupta, P., *The economics of biodiversity: the Dasgupta review: full report* (HM Treasury, 2021)

➤ ***Early use of climate-adapted crops:***
Madella, M. *et al.*, 'Microbotanical Evidence of Domestic Cereals in Africa 7000 Years Ago.' *PLOS ONE* 9 (2014): e110177
doi: 10.1371/journal.pone.0110177

Reed, K. and Ryan, P., 'Lessons from the past and the future of food.' *World Archaeology* 51 (2019): 1–16
doi: 10.1080/00438243.2019.1610492

CHAPTER 7

➤ ***Bible quotes:***
Mclaughlin, R. P., 'A Meatless Dominion: Genesis 1 and the Ideal of Vegetarianism.' *Biblical Theology Bulletin* 47 (2017): 144–154
doi: 10.1177/0146107917715587

➤ ***Whale decomposition:***
Glover, A., 'What happens when whales die?' *NHM - What on Earth?*
Available at: www.nhm.ac.uk/discover/what-happens-when-whales-die.html

➤ ***Wolf re-introduction to Yellowstone:***
Farquhar, B., 'Wolf Reintroduction Changes Yellowstone Ecosystem.' *Yellowstone National Park* (30 June 2021)
Available at: www.yellowstonepark.com/things-to-do/wildlife/wolf-reintroduction-changes-ecosystem/

Peglar, T., 'What Happened to Yellowstone's Wolves After Reintroduction in 1995?' *Yellowstone National Park* (30 June 2021) Available at: www.yellowstonepark.com/park/conservation/yellowstone-wolves-reintroduction/

Smith, D.W., Stahler, D.R., MacNulty, D.R. (eds.), *Yellowstone Wolves: Science and Discovery in the World's First National Park* (Chicago University Press, 2020).

➤ ***Keystone species:***

Biologydictionary.net Editors, 'Keystone Species – Definition and Examples.' *Biology Dictionary* (25 December 2017) Available at: biologydictionary.net/keystone-species/

BirdNote and McCann, M., 'Woodpeckers as Keystone Species.' *Audubon* (20 August 2013)
Available at: www.audubon.org/news/woodpeckers-keystone-species

➤ ***Hawaiian lobelioids:***

Antonelli, A., 'Have giant lobelias evolved several times independently? Life form shifts and historical biogeography of the cosmopolitan and highly diverse subfamily Lobelioideae (Campanulaceae).' *BMC Biology* 7 (2009): 82
doi: 10.1186/1741-7007-7-82

Givnish, T. J. *et al.*, 'Origin, adaptive radiation and diversification of the Hawaiian lobeliads (Asterales: Campanulaceae).' *Proceedings of the Royal Society B: Biological Sciences* 276 (2009): 407–416
doi: 10.1098/rspb.2008.1204

CHAPTER 8

➤ ***Op-ed in* Washington Post*:***
Antonelli, A. and Perrigo, A., 'Opinion | We must protect biodiversity.' *Washington Post* (15 December 2017).
Available at: www.washingtonpost.com/opinions/2017/12/15/53e6147c-e0f7-11e7-b2e9-8c636f076c76_story.html

Antonelli, A. and Perrigo, A., 'The science and ethics of extinction.' *Nature Ecology & Evolution* 2 (2018): 581
doi: 10.1038/s41559-018-0500-z

Pyron, R. A., 'Perspective | We don't need to save endangered species. Extinction is part of evolution.' *Washington Post* (22 November 2017).
Available at: www.washingtonpost.com/outlook/we-dont-need-to-save-endangered-species-extinction-is-part-of-evolution/2017/11/21/57fc5658-cdb4-11e7-a1a3-0d1e45a6de3d_story.html

➤ ***Rights of nature:***
GARN (Global Alliance for the Rights of Nature), 'What is Rights of Nature?' *rightsofnature.org*
Available at: www.therightsofnature.org/what-is-rights-of-nature/

'Rights of nature.' *United Nations* (2021)
Available at: http://www.harmonywithnatureun.org/rightsOfNaturePolicies/

➤ ***Blog on Amazon fires:***
Antonelli, A., 'The Amazon is burning. Will the world just watch?' *Kew Read & Watch* (23 August 2019)
Available at: www.kew.org/read-and-watch/amazon-fires-brazil

CHAPTER 9

➤ ***Effects of habitat loss and intensification of land use on biodiversity:***

Díaz, S. *et al.*, 'Pervasive human-driven decline of life on Earth points to the need for transformative change.' *Science* 366 (2019): 1327
doi: 10.1126/science.aax3100

Ellis, E. C. *et al.*, 'People have shaped most of terrestrial nature for at least 12,000 years.' *Proceedings of the National Academy of Sciences* 118 (2021): e2023483118
doi: 10.1073/pnas.2023483118

Godfray, H. C. J. *et al.*, 'Meat consumption, health, and the environment.' *Science* 361 (2018): 243
doi: 10.1126/science.aam5324

➤ ***The Great Acceleration and its slowdown:***

Dorling, D. *Slowdown: The end of the Great Acceleration – and Why It's Good for the Planet, the Economy, and Our Lives*. (Yale University Press, 2020).

➤ ***Devils Hole pupfish:***

NatureServe, 'IUCN Red List of Threatened Species: Cyprinodon diabolis.' *IUCN Red List of Threatened Species* (2014)
Available at: www.iucnredlist.org/species/6149/15362335

➤ ***Insect declines:***

Hallmann, C. A. *et al.*, 'More than 75 percent decline over 27 years in total flying insect biomass in protected areas.' *PLOS ONE* 12 (2017): e0185809
doi: 10.1371/journal.pone.0185809

McCarthy, M., *The moth snowstorm: nature and joy* (John Murray, 2015)

Seibold, S. *et al.*, 'Arthropod decline in grasslands and forests is associated with landscape-level drivers' *Nature* 574 (2019): 671–674
doi: 10.1038/s41586-019-1684-3

➤ ***Loss of wetlands:***

Davidson, N. C., 'How much wetland has the world lost? Long-term and recent trends in global wetland area.' *Marine and Freshwater Research* 65 (2014): 934–941
doi: 10.1071/MF14173

➤ ***Madagascar's grasslands:***

Solofondranohatra, C. L. *et al.*, 'Fire and grazing determined grasslands of central Madagascar represent ancient assemblages.' *Proceedings of the Royal Society B: Biological Sciences* 287 (2020): 20200598
doi: 10.1098/rspb.2020.0598

CHAPTER 10

➤ ***Reptile pet trading:***

Marshall, B. M., Strine, C. and Hughes, A. C., 'Thousands of reptile species threatened by under-regulated global trade.' *Nature Communications* 11 (2020): 4738.
doi: 10.1038/s41467-020-18523-4

➤ ***Timber industry and identification:***

Meier, E., 'Restricted and Endangered Wood Species.' *The Wood Database*
Available at: www.wood-database.com/wood-articles/restricted-and-endangered-wood-species/

World Forest ID:
worldforestid.org

World Bank, 'Forests Generate Jobs and Incomes.' *worldbank.org* (16 March 2016)
Available at: www.worldbank.org/en/topic/forests/brief/forests-generate-jobs-and-incomes

WWF (World Wildlife Fund), 'Responsible Forestry | Timber.' *worldwildlife.org*
Available at: www.worldwildlife.org/industries/timber

➤ ***Elephant tusks and decline:***

Barnes, R.F.W., 'Is there a future for elephants in West Africa?'. *Mammal Review* 29(3): 175–200. (04 January 2002).
doi: 10.1046/j.1365-2907.1999.00044.x

Flamingh, A. de *et al.*, 'Sourcing Elephant Ivory from a Sixteenth-Century Portuguese Shipwreck.' *Current Biology* 31 (2021): 621-628.e4
doi: 10.1016/j.cub.2020.10.086

Sayol, F. *et al.*, 'Anthropogenic extinctions conceal widespread evolution of flightlessness in birds'. *Science Advances* 6 (49). (2020)
Doi: 10.1126/sciadv.abb6095

Temming, M., 'Ivory from a 16th century shipwreck reveals new details about African elephants.' *Science News* (17 December 2020)
Available at: www.sciencenews.org/article/ivory-shipwreck-african-elephants-tusk-dna-bom-jesus

CHAPTER 11

➤ ***Changes in precipitation:***
Hausfather, Z., 'Explainer: What climate models tell us about future rainfall.' *Carbon Brief* (19 January 2018)
Available at: www.carbonbrief.org/explainer-what-climate-models-tell-us-about-future-rainfall

➤ ***Sources of greenhouse gases:***
C2ES (Center for Climate and Energy Solutions), 'Global Emissions.' *c2es.org* (6 January 2020)
Available at: www.c2es.org/content/international-emissions/

➤ ***Human climatic niche:***
Gorvett, Z., 'The never-ending battle over the best office temperature.' *BBC Worklife* (20 June 2016)
Available at: www.bbc.com/worklife/article/20160617-the-never-ending-battle-over-the-best-office-temperature

Xu, C. *et al.*, 'Future of the human climate niche.' *Proceedings of the National Academy of Sciences* 117 (2020): 11350–11355
doi: 10.1073/pnas.1910114117

➤ ***Adaptation or dispersal to survive climate change:***
Quintero, I. and Wiens, J. J., 'Rates of projected climate change dramatically exceed past rates of climatic niche evolution among vertebrate species.' *Ecology Letters* 16 (2013): 1095–1103.
doi: 10.1111/ele.12144

Román-Palacios, C. and Wiens, J. J., 'Recent responses to climate change reveal the drivers of species extinction and survival.' *Proceedings of the National Academy of Sciences* 117 (2020): 4211–4217
doi: 10.1073/pnas.1913007117

➤ ***Mountains, climate and biodiversity:***
Hoorn, C., Perrigo, A. and Antonelli, A. (eds.), *Mountains, Climate and Biodiversity* (Wiley Blackwell, 2018)

Perrigo, A., Hoorn, C. and Antonelli, A., 'Why mountains matter for biodiversity.' *Journal of Biogeography* 47 (2020): 315–325
doi: 10.1111/jbi.13731

➤ ***Documented shifts in species ranges:***
Morueta-Holme, N. *et al.*, 'Strong upslope shifts in Chimborazo's vegetation over two centuries since Humboldt.' *Proceedings of the National Academy of Sciences* 112 (2015): 12741–12745
doi: 10.1073/pnas.1509938112 (See also two letters linked to the online article, by Feeley and Rehm [2015] and by Sklenář [2016] that further discuss and help refine this paper's findings.)

Parmesan, C. *et al.*, 'Poleward shifts in geographical ranges of butterfly species associated with regional warming.' *Nature* 399 (1999): 579–583
doi: 10.1038/21181

Parmesan, C. and Yohe, G., 'A globally coherent fingerprint of climate change impacts across natural systems.' *Nature* 421 (2003): 37–42
doi: 10.1038/nature01286

➤ ***Phenological changes:***
BBC, 'Japan's cherry blossom "earliest peak since 812".' *BBC News* (30 March 2021)
Available at: www.bbc.com/news/world-asia-56574142

➤ ***Impact on polar species:***
WWF (World Wildlife Fund), '11 Arctic species affected by climate change.' *wwf.org*
Available at: www.wwf.org.uk/updates/11-arctic-species-affected-climate-change

➤ ***Coral reefs and climate change:***
IUCN (International Union for Conservation of Nature), 'Coral reefs and climate change.' *IUCN Issues Brief* (6 November 2017)
Available at: www.iucn.org/resources/issues-briefs/coral-reefs-and-climate-change

➤ ***1.5 vs 2.0 degree difference:***
Thompson, A., 'What's in a Half a Degree? 2 Very Different Future Climates.' *Scientific American* (17 October 2018)
Available at: www.scientificamerican.com/article/whats-in-a-half-a-degree-2-very-different-future-climates/

Warren, R. *et al.*, 'The projected effect on insects, vertebrates, and plants of limiting global warming to 1.5°C rather than 2°C.' *Science* 360 (2018): 791–795
doi: 10.1126/science.aar3646

➤ ***Carbon emissions and ocean acidification:***
Doney, S. C. *et al.*, 'Ocean Acidification: The Other CO2 Problem.' *Annual Review of Marine Science* 1 (2009): 169–192
doi: 10.1146/annurev.marine.010908.163834
Dupont, S. and Pörtner, H., 'Get ready for ocean acidification.' *Nature* 498 (2013): 429–429
doi: 10.1038/498429a

'Global CO2-emissions.' *The World Counts*
Available at: www.theworldcounts.com/challenges/climate-change/global-warming/global-co2-emissions/story

➤ ***Extreme weather:***

Leslie, T., Byrd, J. and Hoad, N., 'See how global warming has changed the world since your childhood.' *ABC News* (5 December 2019)
Available at: www.abc.net.au/news/2019-12-06/how-climate-change-has-impacted-your-life/11766018

➤ ***Australian bushfires:***

BBC, 'Australian bush fires: Royal Botanic Gardens storing seeds.' *BBC News* (7 February 2020)
Available at: www.bbc.com/news/uk-england-51414320

Gutiérrez, P. *et al.*, 'How fires have spread to previously untouched parts of the world.' *The Guardian* (19 February 2021)
Available at: www.theguardian.com/environment/ng-interactive/2021/feb/19/how-fires-have-spread-to-previously-untouched-parts-of-the-world

Readfearn, G. and Morton, A., 'Almost 3 billion animals affected by Australian bushfires, report shows.' *The Guardian* (28 July 2020)
Available at: www.theguardian.com/environment/2020/jul/28/almost-3-billion-animals-affected-by-australian-megafires-report-shows-aoe

CHAPTER 12

➤ ***Invasion meltdown:***

Crego, R. D., Jiménez, J. E. and Rozzi, R., 'A synergistic trio of invasive mammals? Facilitative interactions among beavers, muskrats, and mink at the southern end of the Americas.' *Biological Invasions* 18 (2016): 1923–1938
doi: 10.1007/s10530-016-1135-0

➤ ***Invasive oysters in Sweden:***
Swedish Agency for Marine and Water Management:
www.havochvatten.se/en/start.html

➤ ***Ocean pollution:***
National Geographic Society, 'Marine Pollution.'
nationalgeographic.org (3 July 2019)
Available at: www.nationalgeographic.org/encyclopedia/marine-pollution/

➤ ***Plastics in seabirds:***
Briggs, H., 'Plastic pollution: "Hidden" chemicals build up in seabirds.' *BBC News* (31 January 2020)
Available at: www.bbc.com/news/science-environment-51285103

➤ ***Plastics and human health:***
Rasool, F. N. *et al.*, 'Isolation and characterization of human pathogenic multidrug resistant bacteria associated with plastic litter collected in Zanzibar.' *Journal of Hazardous Materials* 405 (2021): 124591
doi: 10.1016/j.jhazmat.2020.124591

Vethaak, A. D. and Legler, J., 'Microplastics and human health.' *Science* 371 (2021): 672–674
doi: 10.1126/science.abe5041

➤ ***Birth-control pills and their effect on fish:***
Kidd, K. A. *et al.*, 'Collapse of a fish population after exposure to a synthetic estrogen.' *Proceedings of the National Academy of Sciences* 104 (2007): 8897–8901
doi: 10.1073/pnas.0609568104

Nikoleris, L., *The estrogen receptor in fish and effects of synthetic estrogens in the environment – Ecological and evolutionary perspectives and societal awareness*. PhD Thesis. (Centre for Environmental and Climate Science (CEC) and Department of Biology, Faculty of Science, Lund University, 2016)

➤ ***Chemical pollution:***

UNEP (United Nations Environment Programme), 'Global Chemicals Outlook II. From Legacies to Innovative Solutions: Implementing the 2030 Agenda for Sustainable Development' *unep.org*
Available at: https://www.unep.org/explore-topics/chemicals-waste/what-we-do/policy-and-governance/global-chemicals-outlook

'The Different Kinds of Chemical Pollution.' *The World Counts*
Available at: www.theworldcounts.com/stories/Chemical_Pollution_Examples

Wang, Z. *et al.*, 'Toward a Global Understanding of Chemical Pollution: A First Comprehensive Analysis of National and Regional Chemical Inventories.' *Environmental Science & Technology* 54 (2020): 2575–2584
doi: 10.1021/acs.est.9b06379

➤ ***Decline of freshwater fish:***

WWF (World Wildlife Fund) *et al.*, *The World's Forgotten Fishes* (WWF, 2021)

➤ ***Light pollution:***

Irwin, A., 'The dark side of light: how artificial lighting is harming the natural world.' *Nature* 553 (2018): 268–270
doi: 10.1038/d41586-018-00665-7

Owens, A. C. S. *et al.*, 'Light pollution is a driver of insect declines.' *Biological Conservation* 241 (2020): 108259
doi: 10.1016/j.biocon.2019.108259

UNEP (United Nations Environment Programme), 'Global light pollution is affecting ecosystems—what can we do?' *unep.org* (13 March 2020)
Available at: www.unep.org/news-and-stories/story/global-light-pollution-affecting-ecosystems-what-can-we-do

➤ ***Noise pollution in the ocean:***

Duarte, C. M. *et al.*, 'The soundscape of the Anthropocene ocean.' *Science* 371 (2021): 583
doi: 10.1126/science.aba4658

➤ ***Infection diseases in wildlife:***

Daszak, P., Cunningham, A. A. and Hyatt, A. D., 'Emerging Infectious Diseases of Wildlife – Threats to Biodiversity and Human Health.' *Science* 287 (2000): 443–449
doi: 10.1126/science.287.5452.443

Grange, Z. L. *et al.*, 'Ranking the risk of animal-to-human spillover for newly discovered viruses.' *Proceedings of the National Academy of Sciences* 118 (2021)
doi: 10.1073/pnas.2002324118

Morand, S. and Lajaunie, C., 'Outbreaks of Vector-Borne and Zoonotic Diseases Are Associated With Changes in Forest Cover and Oil Palm Expansion at Global Scale.' *Frontiers in Veterinary Science* 8 (2021): 661063
doi: 10.3389/fvets.2021.661063

Scheele, B. C. *et al.*, 'Amphibian fungal panzootic causes catastrophic and ongoing loss of biodiversity.' *Science* 363 (2019): pp. 1459–1463
doi: 10.1126/science.aav0379

CHAPTER 13

➤ ***Golden rules or reforestation and declaration:***
Brewer, G., '10 golden rules for restoring forests.' *Kew Read & Watch* (26 January 2021).
Available at: www.kew.org/read-and-watch/10-golden-rules-for-reforestation

The Declaration Drafting Committee, 'Kew declaration on reforestation for biodiversity, carbon capture and livelihoods'. *Plants, People, Planet.* (12 October 2021)
Available at: doi: 10.1002/ppp3.10230

Sacco, A. D. *et al.*, 'Ten golden rules for reforestation to optimize carbon sequestration, biodiversity recovery and livelihood benefits.' *Global Change Biology* 27 (2021): 1328–1348
doi: 10.1111/gcb.15498

➤ ***Reforestation conference:***
Royal Botanic Gardens, Kew, 'Reforestation for Biodiversity, Carbon Capture and Livelihoods Conference.' *kew.org* (24-26 February 2021)
Available at: www.kew.org/science/engage/get-involved/conferences/reforestation-biodiversity-carbon-capture-livelihoods

➤ ***Nature-based solutions:***
Nature-Based Solutions Initiative:
www.naturebasedsolutionsinitiative.org/

➤ ***Halting biodiversity loss:***
Díaz, S. *et al.*, 'Pervasive human-driven decline of life on Earth points to the need for transformative change.' *Science* 366 (2019): 1327
doi: 10.1126/science.aax3100

Leclère, D. *et al.*: 'Bending the curve of terrestrial biodiversity needs an integrated strategy.' *Nature* 585 (2020): 551–556 doi: 10.1038/s41586-020-2705-y.

➤ ***Roads – their total length and biodiversity impact:***

Barber, C. P. *et al.*, 'Roads, deforestation, and the mitigating effect of protected areas in the Amazon.' *Biological Conservation* 177 (2014): 203–209
doi: 10.1016/j.biocon.2014.07.004

Hoff, K. and Marlow, R., 'Impacts of vehicle road traffic on desert tortoise populations with consideration of conservation of tortoise habitat in southern Nevada.' *Chelonian Conservation and Biology* 4 (2002): 449–456

➤ ***Effectiveness of protected areas:***

Geldmann, J. *et al.*, 'A global analysis of management capacity and ecological outcomes in terrestrial protected areas.' *Conservation Letters* 11 (2018): e12434
doi: 10.1111/conl.12434

Geldmann, J. *et al.*, 'A global-level assessment of the effectiveness of protected areas at resisting anthropogenic pressures.' *Proceedings of the National Academy of Sciences* 116 (2019): 23209–23215
doi: 10.1073/pnas.1908221116

Watson, J. E. M. *et al.*, 'The performance and potential of protected areas.' *Nature* 515 (2014): 67–73
doi: 10.1038/nature13947

➤ ***Mapping priority areas for plant conservation:***

Kew Science News, 'Ebo Forest logging plans suspended.' *Kew Read & Watch* (19 August 2020)

Available at www.kew.org/read-and-watch/ebo-forest-logging-suspended

'Tropical Important Plant Areas (TIPAs).' *kew.org*
Available at: www.kew.org/science/our-science/projects/tropical-important-plant-areas

➤ ***Changing views on conservation:***

Mace, G. M., 'Whose conservation?' *Science* 345 (2014): 1558–1560
doi: 10.1126/science.1254704

➤ ***Ecocide:***

Antonelli, A. and Thiel, P., 'Ecocide must be listed alongside genocide as an international crime.' *The Guardian* (22 June 2021).
Available at: www.theguardian.com/environment/commentisfree/2021/jun/22/ecocide-must-be-listed-alongside-genocide-as-an-international-aoe

Stop Ecocide International:
www.stopecocide.earth

➤ ***Perfluoroalkyl Substances (PFAS):***

EPA (United States Environmental Protection Agency), 'Basic Information on PFAS.' *epa.gov*
Available at: www.epa.gov/pfas/basic-information-pfas

Schrenk, D. *et al.*, 'Risk to human health related to the presence of perfluoroalkyl substances in food.' *EFSA Journal* 18 (2020): e06223
doi: 10.2903/j.efsa.2020.6223

Silva, A. O. D. *et al.*, 'PFAS Exposure Pathways for Humans and Wildlife: A Synthesis of Current Knowledge and Key Gaps in Understanding.' *Environmental Toxicology and Chemistry* 40 (2021): 631–657
doi: 10.1002/etc.4935

➤ ***Future food consumption:***
FAO (Food and Agriculture Organization), *The future of food and agriculture – Trends and challenges* (Food and Agriculture Organization of the United Nations, 2017).

Potter, N., 'Can We Grow More Food in 50 Years Than in All of History?' *ABC News* (2 October 2009)
Available at: abcnews.go.com/Technology/world-hunger-50-years-food-history/story?id=8736358

➤ ***Food waste:***
Depta, L., 'Global Food Waste and its Environmental Impact.' *reset.org* (September 2018)
Available at: en.reset.org/knowledge/global-food-waste-and-its-environmental-impact-09122018

FAO Technical Platform on the Measurement and Reduction of Food Loss and Waste:
www.fao.org/platform-food-loss-waste/en/

USDA (U.S. Department of Agriculture), 'Food Waste FAQs.' *usda.gov*
Available at: www.usda.gov/foodwaste/faqs

➤ ***Biomimetics:***
'Biomimetics.' *Wikipedia* (2021)
Available at: en.wikipedia.org/w/index.php?title=Biomimetics&oldid=1032654369

➤ ***Investments in natural vs produced capital:***
Dasgupta, P., *The economics of biodiversity: the Dasgupta review: full report* (HM Treasury, 2021)

➤ ***Costs of delaying action:***
Vivid Economics and Natural History Museum, *The Urgency of Biodiversity Action* (Natural History Museum, 2021)

➤ ***New Zealand's alternative to economic growth:***
Te Tai Ōhanga – The Treasury, 'Wellbeing Budget 2021: Securing Our Recovery.' *treasury.govt.nz* (2021)
Available at: www.treasury.govt.nz/publications/wellbeing-budget/wellbeing-budget-2021-securing-our-recovery-html

CHAPTER 14

FOOD

➤ ***Environmental and health impact of meat consumption:***
Godfray, H. C. J. *et al.*, 'Meat consumption, health, and the environment.' *Science* 361 (2018): 243
doi: 10.1126/science.aam5324

Mekonnen, M. M. and Hoekstra, A. Y., *The green, blue and grey water footprint of farm animals and animal products* (UNESCO-IHE Institute for Water Education, 2010)

Pimentel, D. and Pimentel, M., 'Sustainability of meat-based and plant-based diets and the environment.' *The American Journal of Clinical Nutrition* 78 (2003): 660S-663S
doi: 10.1093/ajcn/78.3.660S

➤ ***Leakage of antibiotics:***
Chen, N., 'Maps Reveal Extent of China's Antibiotics Pollution.' *News Updates - Chinese Academy of Sciences* (15 July 2015)
Available at: english.cas.cn/newsroom/archive/news_archive/nu2015/201507/t20150715_150362.shtml

Tiseo, K. *et al.*, 'Global Trends in Antimicrobial Use in Food Animals from 2017 to 2030.' *Antibiotics* 9 (2020): 918
doi: 10.3390/antibiotics9120918

➤ ***Eating insects:***

van Huis, A. *et al.*, *Edible insects: future prospects for food and feed security* (Food and Agriculture Organization of the United Nations, 2013)

➤ ***Untapped plant and fungal diversity:***

Antonelli, A. *et al.*, *State of the World's Plants and Fungi 2020* (Royal Botanic Gardens, Kew, 2020)
doi: 10.34885/172

Royal Botanic Gardens, Kew (ed.), 'Special Issue: Protecting and sustainably using the world's plants and fungi.' *PLANTS, PEOPLE, PLANET* 2 (2020): 367-579
Available at: nph.onlinelibrary.wiley.com/toc/25722611/2020/2/5

➤ ***Mycoprotein / fungi:***

Department of Food Science University of Copenhagen (UCPH FOOD), 'Growing sustainable oyster mushrooms on by-products.' *Department of Food Science News* (2 July 2020)
Available at: food.ku.dk/english/news/2020/growing-sustainable-oyster-mushrooms-on-by-products/

Souza Filho, P. F. *et al.*, 'Mycoprotein: environmental impact and health aspects.' *World Journal of Microbiology and Biotechnology* 35 (2019): 147
doi: 10.1007/s11274-019-2723-9

➤ ***Algal production:***

Sudhakar, M. P. and Viswanaathan, S. 'Algae as a Sustainable and Renewable Bioresource for Bio-Fuel Production' in Singh,

J. S. and Singh, D. P. (eds) *New and Future Developments in Microbial Biotechnology and Bioengineering* (Elsevier, 2019), pp. 77–84
doi: 10.1016/B978-0-444-64191-5.00006-7

➤ ***Wasting food:***

Freier, A., 'Pity the Ugly Carrot – It Could Reduce Our Food Waste.' *Medium* (27 September 2019)
Available at: anne-f.medium.com/pity-the-ugly-carrot-it-could-reduce-our-food-waste-ed8fb037ba36

➤ ***Corporate initiatives towards sustainability:***

Sustainable Markets Initiative:
www.sustainable-markets.org/home

United Nations Global Compact:
https://www.unglobalcompact.org/

AT HOME

➤ ***Cotton vs other sources of fibre:***

Cherrett, N. *et al.*, *Ecological footprint and water analysis of cotton, hemp and polyester. Report prepared for and reviewed by BioRegional Development Group and World Wide Fund for Nature – Cymru* (Stockholm Environment Institute, 2005)

Rana, S. *et al.*, 'Carbon Footprint of Textile and Clothing Products' in Muthu, S. S. (ed.) *Handbook of Sustainable Apparel Production* (CRC Press, 2015)
doi: 10.1201/b18428

The Ettitude Team, 'Rayon, Modal, Lyocell – Who's the Fairest of Them All?' *ettitude journal* (1 July 2017)
Available at: ettitude.com/impact/whos-the-fairest-of-them-all

➤ ***Wood from endangered trees:***
BGCI (Botanic Gardens Conservation International), 'ThreatSearch.' *bgci.org*
Available at: tools.bgci.org/threat_search.php

Fauna & Flora International and BGCI (Botanic Gardens Conservation International), 'Global Trees Campaign.' *globaltrees.org*
Available at: globaltrees.org

Meier, E., 'Restricted and Endangered Wood Species.' *The Wood Database*
Available at: www.wood-database.com/wood-articles/restricted-and-endangered-wood-species/

➤ ***Cleaning:***
BH&G Editors, 'How to Clean Almost Every Surface of Your Home With Vinegar.' *Better Homes & Gardens* (10 March 2020)
Available at: www.bhg.com/homekeeping/house-cleaning/tips/cleaning-with-vinegar/

➤ ***Cosmetics:***
Botanical Trader, 'Are Cosmetics Bad For The Environment?' *Botanical Trader* (20 January 2019)
Available at: botanicaltrader.com/blogs/news/how-your-beauty-products-are-killing-coral-reefs-turtles-rainforests-more

➤ ***Family planning for climate:***
Lunds University, 'The four lifestyle choices that most reduce your carbon footprint.' *Lunds University News* (12 July 2017)
Available at: www.lunduniversity.lu.se/article/four-lifestyle-choices-most-reduce-your-carbon-footprint

Wynes, S. and Nicholas, K. A., 'The climate mitigation gap: education and government recommendations miss the most

effective individual actions.' *Environmental Research Letters* 12 (2017): 074024
doi: 10.1088/1748-9326/aa7541

➤ ***Cooking energy:***
Hager, T. J. and Morawicki, R., 'Energy consumption during cooking in the residential sector of developed nations: a review.' *Food Policy* 40 (2013): 54–63
doi: 10.1016/j.foodpol.2013.02.003

➤ ***Solar energy:***
Evans, S., 'Solar is now "cheapest electricity in history", confirms IEA.' *Carbon Brief* (13 October 2020).
Available at: www.carbonbrief.org/solar-is-now-cheapest-electricity-in-history-confirms-iea

➤ ***The impact of cats on biodiversity:***
Farmer, C. and Sizemore, G., 'For Rare Hawaiian Birds, Cats Are Unwelcome Neighbors.' *Birdcalls – News and Perspectives on Bird Conservation* (27 February 2016).
Available at: abcbirds.org/for-rare-hawaiian-birds-cats-unwelcome/

Hawaii Invasive Species Council, 'Feral Cats.' *dlnr.hawaii.gov* (21 January 2016)
Available at: dlnr.hawaii.gov/hisc/info/invasive-species-profiles/feral-cats/

Loss, S. R., Will, T. and Marra, P. P., 'The impact of free-ranging domestic cats on wildlife of the United States.' *Nature Communications* 4 (2013): 1396
doi: 10.1038/ncomms2380

Medina, F. M. *et al.*, 'A global review of the impacts of invasive cats on island endangered vertebrates.' *Global Change Biology* 17 (2011): 3503–3510
doi: 10.1111/j.1365-2486.2011.02464.x

Platt, J. R., 'Hawaii's Invasive Predator Catastrophe.' *The Revelator* (24 June 2020)
Available at: therevelator.org/hawaii-predator-catastrophe/

➤ ***Environmental impacts of food consumption by dogs and cats:***

Okin, G. S., 'Environmental impacts of food consumption by dogs and cats.' *PLOS ONE* 12: e0181301 (2017)
doi: 10.1371/journal.pone.0181301

➤ ***Dog vs hamster:***

Power, J., 'How big is your pet's environmental paw-print?' *The Sydney Morning Herald* (1 September 2019)
Available at: www.smh.com.au/environment/sustainability/how-big-is-your-pet-s-environmental-paw-print-20190830-p52mbz.html

➤ ***Challenges with vegetarian alternatives to cats and dogs:***

Dowling, S., 'Can you feed cats and dogs a vegan diet?' *BBC Future* (4 March 2020)
Available at: www.bbc.com/future/article/20200304-can-you-feed-cats-and-dogs-a-vegan-diet

OUR BACKYARD

➤ ***People living in urban areas:***

Ritchie, H. and Roser, M., 'Urbanization.' *Our World in Data* (2018)
Available at: ourworldindata.org/urbanization.

➤ ***Role of gardens for biodiversity:***

Gaston, K. J. *et al.*, 'Urban domestic gardens (II): experimental tests of methods for increasing biodiversity.' *Biodiversity & Conservation* 14 (2005): 395
doi: 10.1007/s10531-004-6066-x

TRANSPORTATION

➤ ***Carbon footprint of different ways of travelling:***

Ritchie, H., 'Which form of transport has the smallest carbon footprint?' *Our World in Data* (2020)
Available at: ourworldindata.org/travel-carbon-footprint.

➤ ***Air pollution from cars:***

EPA (United States Environmental Protection Agency), 'Research on Health Effects, Exposure, & Risk from Mobile Source Pollution.' *epa.gov* (7 December 2016)
Available at: www.epa.gov/mobile-source-pollution/research-health-effects-exposure-risk-mobile-source-pollution

➤ ***Commuting distances:***

Textor, C., 'Average distance travelled for commuting purposes in China in 2020, by city size (in kilometers)', *Statista* (10 June 2020)
Available at: www.statista.com/statistics/1121851/china-average-commute-distance-by-city-size/

SOFT POWER

➤ ***Carbon emissions from UK military:***

Parkinson, S. and SGR (Scientists for Global Responsibility), *The Environmental Impacts of the UK Military Sector* (Scientists for Global Responsibility & Declassified UK, 2020)

➤ ***Reducing the climatic impact of companies:***

Carbon Literacy:
https://carbonliteracy.com/

Carbon Offset Guide:
www.offsetguide.org

Race To Zero:
https://unfccc.int/climate-action/race-to-zero-campaign

Science Based Targets:
sciencebasedtargets.org

Sustainability at the workplace:
https://www.wwf.org.uk/sites/default/files/2020-08/WWF UK Sustainable Office Guide 2020.pdf
https://juliesbicycle.com/resources-green-office-guide-2015/

➤ ***Progress and challenges in labelling products:***
Perrigo, A. *et al.*, 'The full impact of supermarket products.' *Springer Nature Sustainability Community* (16 July 2020) Available at: sustainabilitycommunity.springernature.com/posts/the-full-impact-of-supermarket-products

INVESTMENTS AND OTHER ACTIONS

➤ ***Elephant poaching:***
Gill, V., 'Extinction: Elephants driven to the brink by poaching.' *BBC News* (25 March 2021). Available at: www.bbc.com/news/science-environment-56510593

➤ ***Donations proportion in the US:***
Charity Navigator, 'Giving Statistics.' *charitynavigator.org* (2018) Available at: www.charitynavigator.org/index.cfm?bay=content.view&cpid=42

➤ ***Problems of investing too much in iconic animals:***
Bee, S., 'F*ck The Pandas: Ugly Animals Deserve Your Attention Too.' *Full Frontal with Samantha Bee* (28 January 2021) Available at: www.youtube.com/watch?app=desktop&v=fy4IhJrSJT4&feature=youtu.be#dialog

➤ ***How pensions are used:***

Mustoe, H., 'What's your pension invested in?' *BBC News* (7 March 2021).
Available at: www.bbc.com/news/business-56170726

Simon, E., 'Majority of workers don't know where pension funds invested.' *Corporate Adviser* (18 December 2019).
Available at: corporate-adviser.com/majority-of-workers-dont-know-where-pension-funds-invested/

EPILOGUE

➤ ***Magnitude of human-driven mammal extinctions:***

Andermann, T. *et al.*, 'The past and future human impact on mammalian diversity.' *Science Advances* 6 (2020): eabb2313
doi: 10.1126/sciadv.abb2313

➤ ***Olof Palme's speech:***

Palme, O., 'Statement by Prime Minister Olof Palme in the Plenary Meeting, June, 6, 1972' in *UN Conference on the Human Environment* (Swedish Delegation to the UN Conference on the Human Environment, 1972)

The speech on video:
https://www.youtube.com/watch?v=0dGIsMEQYgI

➤ ***Outcomes towards meeting the 20 Aitchi Biodiversity Targets:***

Díaz, S. *et al.*, 'Pervasive human-driven decline of life on Earth points to the need for transformative change.' *Science* 366 (2019): 1327
doi: 10.1126/science.aax3100

Secretariat of the Convention on Biological Diversity, *Global Biodiversity Outlook 5* (Montreal, 2020)

➤ ***Global forest loss:***
Global Forest Watch:
www.globalforestwatch.org

Global Forest Watch, 'Global Forest Watch Dashboard.' *gfw. global*
Available at: gfw.global/3iWrd5p.

➤ ***The State of the World's Forests in 2020:***
FAO (Food and Agriculture Organization) and UNEP (United Nations Environment Programme), *The State of the World's Forests 2020.* (FAO and UNEP, 2020)
doi: 10.4060/ca8642en

➤ ***Sustainable development goals:***
United Nations, 'The 17 Sustainable Development Goals.' *sdgs. un.org* (2015)
Available at: sdgs.un.org/goals

➤ ***Making peace with nature:***
UNEP (United Nations Environment Programme), *Making Peace with Nature: A scientific blueprint to tackle the climate, biodiversity and pollution emergencies* (Nairobi, 2021)

PICTURE CREDITS

p.10 A herbarium specimen deposited at the Royal Botanic Gardens, Kew – © Board of Trustees, RBG Kew

p.18 The Hubble Sphere and the Tree of Life – © Meghan Spetch

p.28 The Biodiversity Star – © Meghan Spetch

p.41 The Theory of Island Biogeography– © Meghan Spetch

p.47 Arabica coffee *Coffea arabica* with its flowers, fruits and beans – © Lizzie Harper

p.51 An artefact made of leaves from the date palm – © Erin Messenger, RBG Kew

p.55 Thylacine *Thylacinus cynocephalus* – © Lizzie Harper

p.57 Charles Darwin's hand drawing of an evolutionary tree – © Cambridge University Library

p.59 Examples of evolutionary trees (phylogenies) – © Meghan Spetch

p.66 Cross-section of the zonation of the underwater community of a British rocky shore – © Lizzie Harper

p.71 Wallace's Zoogeographic regions – © Meghan Spetch

p.78 Orchid *Cycnoches guttulatum* with Orchid bee *Euglossa cybelia* – © Lizzie Harper

p.83 Drawings of two species in the plant genus *Ciliosemina* – © Kirsten Tind and Olof Helje

p.95 Rewilding: Wolf, *Canis lupus*, hunting Elk, *Cervus canadensis* in Yellowstone National Park– © Lizzie Harper

p.106 Major threats to biodiversity – © Meghan Spetch

p.108 Scarlet macaw *Ara macao* – © Lizzie Harper

p.119 Cretan orchid *Cephalanthera cucullata* – © Lizzie Harper

p.123 Cross section of a wood sample (the common oak) – © Peter Gasson, RBG Kew
p.132 Coral reef seascape – © Lizzie Harper
p.139 Beavers at work in the world's southernmost inhabited island, Isla Navarino
p.150 Bending the curve of biodiversity loss – © Meghan Spetch
p.154 Oak tree *Quercus robur* – © Lizzie Harper
p.159 Protecting biodiversity and its benefits to people – © Mauricio Diazgranados Cadelo, RBG Kew
p.179 Wonky fruit and vegetables – © Lizzie Harper
p.181 Oyster mushrooms *Pleurotus ostreatus* – © Lizzie Harper
p.184 Hemp cloth string and plant *Cannabis sativa* – © Lizzie Harper
p.194 Hedgehog *Erinaceus europaeus* with nettles, log pile, fungi and Small tortoiseshell butterfly *Aglais urticae* – © Lizzie Harper
p.203 Disclosing the socio-environmental impact of a product – © Meghan Spetch
p.214 Hidden Universe: Starry sky – © Lizzie Harper

INDEX

Note: page numbers in **bold** refer to information contained in captions. Page numbers preceded by an 'n' refer to information contained within footnotes.

acid rain 142
adaptation 96, 115, 129, 136, 145
Africa 2, 39, 69, 73, 87, n 111, 115, 117, 155, 175, 204–5
  eastern 3, 23, 89–90
  Northern 51
  southern 94, 124
  tropical 112
  West 124
agriculture
  and climate change 128
  freshwater use 111, 113
  intensification 165–6
  slash and burn practices 102–4, 110, 117
  traditional practices 166
  see also farmers
agrobiodiversity 166–7
algae 132–3, 180
Amazon rainforest 7, 20, 23–5, 38, 82, 160–1
  and deforestation 102–4, 158, 166, 199
  and pollution 142
  and species rarity 43
  tipping point 24–5, 75
ancestors 2–4, 18, **18**
Andermann, Tobias 210
Andersson, Lennart 81–2
Andes 7, 13, 69, 84, 130, **159**
animal feed 111, 175, 176, 192
Antarctica 40, 162
anthropocentrism 99
antibiotic resistance 176
Antonelli Lab viii
Aono, Yasuyuki 135
aquatic environments 142
  see also oceans
Arakaki, Mónica 73–4
Aral Sea 75
Archaea 17, 20
Arctic 35–6, 46, 62–3, 97
Aristotle 4
ash dieback disease 48–9
Asia 23, 48, 69, 74, 114, 118, 138
  East 87, 147
  Middle 87
  northern 72
  South 52, 162
  Southeast 112, 122
  western 51
asteroid collisions 24, 44, 151
Atlantic Forest 38
Australia 23, 40, 55, 73, 87, 115, 128, 131, 135–6, 146, 165
Austria 199
avocado 178–80

bacteria 17, 20, 38, 145, 146
Baker, William 121–2
banana crop 87
Bangladesh 102, 183
Banks, Joseph 7
Barro Colorado Island 97
Batrachochytrium 146
bats 33, 145
Batty, Wilf 54
Beagle (ship) 8
beavers 93, 137–8, **139**
beech 57–8, 137
beef 102, 111, 174–5
bees 37, 88
Bhutan 170
Bible 91
biodiversity ix–xi, 2–3, 16–17
  complexity 29, 209

crisis x, 151
estimation 15–22
loss 76
bending the curve of **150**
costs of 170
drivers of *see* biodiversity threats
and food production 174
future of 210, 211
and gene-editing technology 46
and human activity 23–5
reversal 25, 26, 157, 165, 167, 169, 212
saving 149–213, **150**
future of 209–13
individual action 171–2, 173–208
large-scale solutions 153–72
nature-based solutions 156
threats to 105–47, **106**, 169
climate change 127–36, 147
emerging diseases 145–7
exploitation 117–26, 147
habitat loss **106**, 109–16, 147
pollution 140–4, 147
value of 77–104
for itself 99–104
for nature 91–8
for us 81–90, 97–8
biodiversity science 100
biodiversity star **28**, 29
ecosystem diversity **28**, 29, 69–76, 87–8, 114
evolutionary diversity **28**, 29, 54–60
functional diversity **28**, 29, 61–8, **66**, 97, 122–5
genetic diversity **28**, 29, 44–53, 76, 114, 122–5
species diversity **28**, 29, 31–43, 122, 136
biogeography ix, 40–2, **41**
biological collections 50–2, **51**
biological corridors 131, 157
biomes 71
biomimetics 168–9
biota 135–6
blight 87, 146–7
Bolivia 84, 102, 159
Bolsonaro, Jair 102–3
Borneo 88
brain size 3
Brassicaceae 85, 86
Brazil vii, ix, 22, 38, 60, 102–4, 109–10, 155, 174, 176, 209
Breman, Elinor 53
bromeliads 38
Buggs, Richard 48
bushfires 135

California 46, 72, 94
Cameroon 160
Canada 69, 72, 165
Canales, Nataly 84
Cañari 84
capital
natural 169–70
produced 169
capybara 109
carbon capture 65, 74, **150**
carbon dioxide emissions 23, 116, 128, 133–4, 136, **150**, 177, 183
offsetting 201
carbon footprints 183, 204
carbon sequestration **150**
Caribbean 66–8, 69
carrot family 5
cars 199–200
Carvalho, Monica 35
cats 6, 96, 125, 191–2
cattle 110–11, 117, 176, 183
see also beef
Central America 73, 97
Cephalanthera cucullata 118–20, **119**
Cerrado savannah 72
charcoal 110
charitable donations 204–6
cherry trees 135
chestnut, American 146–7
chicken 174–5
Chile 7, 72, 137, 178
Chimborazo volcano 69
Chimú 84
China 4, 112, 118, 176, 198
chlorofluorocarbons (CFCs) 162
chloroplasts 65
Chomelia 43

Ciliosemina 82, **83**
Cinchoneae 82–4
circular production 168
classification 5–7, 8, 21, n 70
cleaning products 188–9
climate change **106**
  and Arctic vegetation 62–3
  as biodiversity threat 127–36
  and coffee production 45–6
  combating **150**
  costs of 170
  denial 202–4
  and ecosystems 71–5
  and re-forestation 64–5
  resilience to 211
  species adaptation to 44–6
climate emergency x
climatic tolerance 46, 128, 130–3
clothing industry 183–5
coconut, double 9, 118
coelacanth **59**
coffee 45–6, **47**, **83**
Coleman, Edith 36–7
colonisation 87, 114
communities 70
compost bins 196
Congo 89
conservation 52, 65, **150**, 157, 159–65
consumption 107, **150**, 175, 185–6
COOP chain 202
coral 132–3, **132**, 156–7
  bleaching 133, 135
Cordiera montana 13
coronaviruses 145
  see also Covid-19 pandemic
cosmetics 188–9
Costa Rica 163
cotton 183–5
Covid-19 88, n 111, 145, 187, 191, 198–9
crabs 15, **66**
  horseshoe 88, 90
Crete 52, 118, **119**
crime, international 164–5
crop wild relatives 166
crops 51–2, 86, 86–7
  diversification 89–90, 166
  genetically modified 46, 165
  and habitat loss 110, 111
  irrigation 113
  monocultures 110, 111, 154
  orphan 166
cyanobacteria 142
cycling 198

dairy products 177
'dark energy' 21, 25–6
'dark matter' 21, 151
Darwin, Charles 7–8, 18, 36, 49, 56, **57**, 124
Dasgupta, Partha 169
date palms 51–2, **51**
David, Aaron 45
de Mestral, George 169
dead zones 142
Death Valley National Park 113
deception strategies 36–7
deforestation 23–5, 45, 160, 186
  and the Amazon 102–4, 158, 166, 199
  controlling 163–4, 166–7
  drivers 111, 112
  and exploitation 117
  global rates of 210
  and palm oil 202
  and roads 158, 197
Deklerck, Victor 121
Demissew, Sebsebe 166
Denmark 167
deoxyribonucleic acid (DNA) 101
  biological collections of 50–2, **51**
  degradation/fragmentation 50–1
  double helix structure 18
  and evolutionary diversity **59**
  exchange across species 39
  of extinct species 34–5
  and gene-editing technology 46
  and genetic diversity 44
  human 50
  mutation 145
  see also DNA barcoding; DNA sequencing
desertification 75
Devils Hole pupfish 112–13
Di Sacto, Alice 155
Díaz, Sandra 171

Diazgranados, Mauricio 178
DiCaprio, Leonardo 160
dichloro-diphenyl-trichloroethane (DDT) 142–3, 162
Dickman, Chris 135
diclofenac 162
diet 2–3
  see also plant-based diets
dinosaurs 74
diseases
  emerging 145–7
  see also specific diseases
DNA barcoding 34
DNA sequencing 17–20, 32–5, 48, 51–2, 58, 60, 121–2, **123**, 124
do Céo Pessoa, Maria 42–3
dodo 125
dogs 31–2, 55, 191–2
dolphins 6, 144, 177
Dominican Republic 120
Dupont, Sam 134

eagles 143, 162
Ebo forest 160
echolocation 33
ecocide 164–5
economic issues 48, 53, 79, 89–90, 170–1
ecosystem diversity **28**, 29, 69–76, 87–8, 114
ecosystem engineers 137
ecosystem services 48, 88, 98
ecosystems 7, 69–76
  and climate change **66**, 132
  degradation 110, 116–17, 136, 157–8, 165, 170, 210–11
  destruction 164
  division of terrestrial 70–1
  fragility 48
  fragmentation 129
  and functional diversity 68
  human 20–1
  protection 157, 165
  resilience **66**, 96
  restoration 156, 165, 210–11
  transience 73–5
  transitions between 73
ecotourism 161, 164
Ecuador 13, 42, 69, 84, 102
Edwards, Erika 74
Ehrlich, Paul 171
Einstein, Albert 8
El Niño 49
Elasmodactylus tetensis 14
electric vehicles 199–200
elephants 3, 94, 124–5, 180, 205
elk, Rocky Mountain 92–3, **95**
endangered species 113
energy 189–91
environmental challenges 7, 58
environmental crisis 153
environmental footprints 174, 182, 201
environmental impact 171
Escobar, Oscar Pérez 51
ethanol, renewable 199
Ethiopia 45, 166
ethnobiology 6
Europe 23, 74, 162
evolution 3, 8, 49–50, 54–60, 58, 70–1, 89, 124
  convergent 55–6
evolutionary diversity (phylogenetic diversity) **28**, 29, 54–60
exploitation of nature 90, **106**, 110–11, 117–26, **150**, 158, 169–70
extinction 17
  bird 125
  and climate change 129
  and deforestation 160
  and the domestic cat 96
  and emerging diseases 146
  and exploitation 117, 120, 122, 125
  and functional diversity 68
  and genetic variation 53
  and human activity 23–5, 54–5, **55**, 115, 210, 212
  mass extinction events 24, 151
  as 'natural process' 99–100
  and pollution 142
  and redundancy 68
  reversal 25
  species at risk of 107
  and species identification 34–5
  and species richness 39
  tree 122

extinction debt 114

farmers 45–6, 54–5, 87, 89–90, 92–4, 102–4, 110
Farooq, Harith 13–14
Fiji 200
finches, Galápagos 8
fish 46, **59**, 112–13, 117, 142, 177
Fleming, Alexander 89
flight 161, 199
Flores 3
food
  production 85–6, 127, 165–8, 174–82, **203**
  security 45–6, 53, 60, 116, 167, 211
forest bridges 129
forests
  benefits to people 88
  boreotropical 74–5
  fragmented 114
  reforestation 153–6
  see also deforestation; rainforests
fossils ix, 17, 35, 74
fox
  Arctic 126, 131
  flying 145
France 164, 167, 199
freedom of movement 157–8, 190
French Guiana 43
frogs 146, 197
fruit 178–80, **179**
fruit fly 50
functional diversity **28**, 29, 61–8, **66**, 97, 122–5
fungal diseases 48, 87, 145–7
fungi 16–17, 20, 33–4, 180
furniture 185–7

Galápagos archipelago 8, 49
gardens 192–6, **194**
Gasson, Peter 121
geckos 13–14
gene-editing technology 46–8
genes 43, 44–53, 76, 89, 131
Genesis 91
genetic diversity **28**, 29, 44–53, 76, 114, 122–5
genetically modified crops 46, 165
genome, human 18
George III 4
Germany 43, 144, 167
global warming 44–6, 62–3, 116, 127–36
'good life' concept 170
Gothenburg Botanical Garden viii
government policy 161–5, 171–2, 202–4
GPS tags 97
Gran Canaria 73
Grant, Peter 49
Grant, Rosemary 49
grasses (Poaceae) 86
grasslands 69, 73–4, 115
  see also savannah
grazing 93, 115
Great Acceleration 111
Great Barrier Reef 135
Great Famine 87
greenhouse gas emissions 25, 65, 128, 177
  see also carbon dioxide emissions; methane
Greenland 88
Gross Domestic Product (GDP) 170
Guinea 200
Guinea-Conakry 159
Gustafsson, Lovisa 35–6
gut microbiome 20–1

habitat degradation 205
habitat encroachment 124
habitat loss **106**, 109–16
habitats, untouched/pristine 110
Hardwick, Kate 155
Hawaii 94–6, 178, 192
health 20–1, 46, 167, 171, 174, 176, 198
hemp 183, **184**
herbarium specimens 9–12, **10**
Hispaniola 120
homes 182–92
Homo
  H. erectus 3
  H. floresiensis 3
  H. neanderthalensis 39
  H. sapiens 22–3, 39, 50, 107
Honduras 138

Hooker, William 9
Hoorn, Carina 35
hormones, synthetic 142
horses 39
Hubble, Edwin 1–2, 25
Hubble Sphere, The **18**
Hughes, Alice 118
human activity 110–11, 128, 151, 210, 212
see also exploitation
Humboldt, Alexander von 7, 69, 71–2, 130
hunter-gatherers 3–4
hunting 92, 117–18, 158, 161–2, 204–5
trophy 124
hydroelectric power 190
hydrothermal vents 70
hyphae 34

IKEA 121–2
iNaturalist app 207
income insecurity 45
India 102, 162, 175, 183
indigenous peoples 6, 23, 64, 101–2, 110–11
Indonesia 202
inequality 86, 87, 107
insects 88, 135
decline in 113–14, 144, 177
eating 177
species numbers n 16
interbreeding 31, 35–9
Inuit people 88
invasive species 137–40, **150**, 155
investments 204–8
Ireland 87
irrigation 113
Isla Navarino 137–8, **139**
Island Biogeography, theory of 40–2, **41**
island universes 25
isotopes 122, 124
Ivory Coast 145

jabirus 109
Japan 168, 175, 180
Jaramillo, Carlos 35
jewellery 186
Juglandaceae 85
jungles 69
Jurassic Park (1993) 35

Kenya 180
keystone species 94
kingfisher 168
Korean Peninsula 146
Kyoto 135

land changes 111
land degradation 160–1
language 4, 5
Lapland 61, 62
latitudinal diversity gradient 42
Latnjajaure, Lake 61
lawns 193–5
Le Moine, Rebecka 164
leather 183
Leavitt, Henrietta 1–2
legume family (Fabaceae) 86, 178
lemmings 131
lemurs 112
Levant 3
lichen 17, 48, 62, 63, 132, 154
life forms 7
life-learners 25
ligers 31
light pollution 143–4, 196–7
lighting, outdoor 196–7
Linnaeus, Carl 5–8, 15, 21, n 70
Lissopimpla excelsa 36–7
livestock 92–4, 111
lobeloids 94–6
local communities 161
logging 160–1, 186–7, 204, 207
Lohmann, Lúcia 60
London 7, 10
lungfish **59**

MacArthur, Robert 40
macaw, scarlet **108**, 109
Mace, Georgina 160
macronutrients 85
Madagascar 70, 112, 115–17, 125, 157, 160–1, 166
mahogany 121
Major, Daphne 49

Malay Archipelago 7
Malaysia 202
Maldonado, Carla 84
malnutrition 86
mammals 33, 39, 135
mammoth 35
Manaus 43
mangroves 97, 156–7
Manhammar, Magnus 164
marsupials 56
Mauritius 125
meadows 69, 137, 193
meat 117, 174–8
  lab-grown 178
medicines 53, 84–5, 88–9, 118
Mediterranean 72, 75, 118, 128
mercury 142, 162
methane 128, 177
Mexico 73, 102, 175
microbes 20
micronutrients 85–6
microplastics 141, 185, 188
Middle East 51
migration 45, 129–30, 143–4, 180, 190
Milky Way 1, 17, 25, 196
mining 142, 162, 200
  underwater 112
mink 137–8
Missouri Botanical Garden 14
moas 125
Molau, Ulf 131
monkeys 23, 25, 97, 101
monocultures 110, 111, 154
morama bean 178
morphological features 115
moss 17, 63, 154
Mount Wilson Observatory 1
mountains 130–1
Mozambique 13–14, 159
Muasya, Muthama 60
Muir, John 62
Müller, Paul 142
mushroom 180–1
  honey (*Armillaria*) 34
  oyster 180, **181**
muskrat 138
mutation 58
mycelium 34
nanoplastics 141
Natural History Museum 10, 170
natural resources 23–5, 64, 110–11, 116, 169–70
natural selection 3, 58, 96, 124
nature 79
  humanity's relationship with 88, 101, 160, 169, 171
  Rights of Nature 101–2
  and the value of biodiversity 91–8
nectar 38, 79, 96, 195
Neotropical regions 71, **71**
Nesbitt, Mark 84
net zero 201
Netherlands 8, 35, 164
Nevada, USA 113, 158
New Guinea 55
New Zealand 89, 102, 125, 170
nitrogen 142, 168
nitrous oxides 128, 142
noise pollution 144, 193
Nordhaus, William 171
North America 23, 35, 69, 74, 88, 146–7, 162
Norway 35–6
nuclear power 189–90

oak 57–8
  common European **123**, **154**
oceans
  acidification 133–4
  ecosystems 69–70
  pollution 140–1, 144
orchids 36, 36–8, 118–20, **119**, 187
  lady's slipper 37, 38
  tongue 36–7
Oregon, US 34
organic produce 177, 183
organism groups 33
Oslo 36
ox (zebu) 115
oyster, Japanese 138–40
ozone layer depletion 162

palm oil 112, 189, 202
Palme, Olaf 210
Panama 35
pandan tree 178

pandemics 88–9
  see also Covid-19
Pantanal, Brazil **108**, 109–10
'paper reserves' 158
Papua New Guinea 16
Páramo vegetation, Andes **159**
Paris Agreement 133
parrot, Dominican 120
Paton, Alan 53
penicillin 89
perfluoroalkyl substances (PFAS) 163
Perrigo, Allison 100
Persson, Claes 13
Peru 13, 84
pesticides 111, 144, 165, 168, 180, 183
pets 118, 187, 191–2
phenology 134–5
Philippines 178
phosphorus 142, 168
photosynthesis 65, 74
  C3/C4 74
  Crassulacean acid metabolism (CAM) 74
Phylocode n 6
Pimiento, Catalina 66
pine 58, 154, 155
Pironon, Sam 178
plankton **66**, 141
plant-based diets 175, 177, 180, 196
plastics 140–1, 180, 185, 188
poisonous plants 5
Poland 48
polar bear 97, 131
Polar regions 61, 131
politics 202–4
pollen 35, 36, 38–9, 96, 110
pollination 36–9, 88
pollinators 96, 135
pollution 140–4, 147, 170
  chemical 141–3, 162–3, 168, 188–9
  light 196–7
  noise 144, 193
  reduction **150**
polychlorinated biphenyls (PCBs) 143, 162
population growth 23, 111, n 111, 124, 165
populations 70
pork 111, 174–5
potato crop 87
poultry 111
poverty 86, 116–17, 167, 210
predators **66**
primates 117
protected areas 157–60, 204
pseudocopulation 36–7

Quechua 84
Quesada, Carlos Alvarado 163
quinine 7, 82, 90

rainfall 24, 72, 128, 136
rainforests 38, 72, 97–8, 115
  see also Amazon rainforest
Ralimanana, Hélène 160–1
range, species 112–13
rattan 121–2
re-forestation 64–5
redundancy 65–8
reindeer 131–2
Reptiles 118, 135
restoration strategies **150**
rewilding 93
rhinos 3, 118
Rio de Janeiro herbarium 9
Ritter, Camila Duarte 20
road kill 158
roads 129, 157–8, 197
rocky shores, species of **66**
Rome Statute 1990s 164
Rønsted, Nina 84
rose, dog 5, 6
rosewood 117, 121, 187
Royal Botanic Gardens, Kew vii, xi, 45, 48, 102, 115, 121, 155, 159, 160, 166, 178
  herbarium specimens 9, **10**
  Madagascar Conservation Centre 160
  Millennium Seed Bank, Wakehurst 52–3
  seed collections 136

Sahel region 75
Sami people 62
sampling, destructive methods 15–16

Saqqara basketry, Egypt 51–2, **51**
satellite imagery 112, 160
savannah 24, 69, 72, 74–5, 110
Sayol, Ferran 125
Scandinavia 72, 126, 131
scent 37
Scientific Revolution 4
Scots pine (*Pinus sylvestris*) 42
sea otter 94
sea urchin 94
seabed, and habitat loss 112
seals 88, 90, 97, 141, 144, 162, 177
seasonality 72, 134–5, 178
seed collections 52–3, 136
selenium 86
Seychelles 118
Shinkansen trains 168
Siberian permafrost 35
'silent violence' 112
slash and burn policy 102–4, 110, 117
Smith, Paul 155
Smith, Rhian 155
soft power 200–4
soil sampling 18–20
Solanaceae 85
solar energy 65, 189, 190
Solofondranohatra, Cédrique 115
South Africa 60, 73, 175
South America 23, 43, 69, 72–3, 82, **108**, 110–11, **159**
Soviet Union 75
soybean 102, 110–11, 117, 134, 154, 166
Spain 162
speciation 12–15, 32, 33, 42, 96, 114
species 3, 4–17, 22, 31–43
  climatic tolerance 128
  collections 8–12
  cryptic 36
  DNA identification 18–20, **18**
  and ecosystem diversity 70
  and evolutionary diversity 55–8, **57**, **59**
  exchanging genes across 38–9
  and gene-editing technology 46
  geographical distribution 39–43
  identification 33–5, 39
  invasive 137–40, **150**, 155
  naming 5–7, 8, 10–12, 21–2
  new 12–15, 32, 33, 42, 96, 114
  numbers 15–17
  preservation 52
  rapid disappearance 23–4
  rarity 42–3, 118–20, **119**
  redundancy 65–8
  of rocky shores **66**
  sightings 206–7
  of socio-economic value 53
  tolerances 7, 40
  variation in behaviour 12
  variation in form 12
  variation within 43, 44
  see also specific species
species diversity **28**, 29, 31–43, 122, 136
species richness 39–40, 63, 96–7
species-area relationship 40, 114
Stalin, Joseph 75
stars 1, 22, 209
stomata 65, 74
Strid, Arne 118
subspecies 32
sugars 65, 74
sulphur dioxide 142
Sumerians 4
supermarkets 167, 202
suspension feeders **66**
sustainability 110–11, 168, 177, 200–1, **203**, 205–6, 211
  and deriving value from biodiversity 90
  and indigenous peoples 23
  and production **150**
  and timber 121, 187
Sustainable Development Goals 211
Sustainable Markets Initiative 168
Svalbard 35
Sweden vii, viii, 15, 91–2, 138–40, 153–4, 202
Switzerland viii

tanager family **59**
Tasmania 54–5
Tasmanian tiger/wolf (thylacine) 54–6, **55**
taxation 161, 163–4, 204
taxonomy n 70

Taylor, Charlotte 14–15
technosphere 141
telescopes 17, 18
termites 169
tetrapods **59**
Theophrastus 4, 8
Thiel, Pella 164
threatened species 118–21
Thunberg, Greta 200
Tibetan Plateau 69
ticks 91–2
tigon 31
timber 79, 120–2, 155, 158, 160, 185
  identification 121–2, **123**
  illegal trade 122, **123**
  see also logging
tipping points 24–5, 75
Tjärnö 15
toads 146
tortoise 125, 158, 187
tourism 117, 161, 164
traditional Chinese medicine 4
traits 63
  see also functional diversity
transportation 197–200
Tree of Life (evolutionary tree/phylogenic tree/phylogeny) **18**, 55–60, **57**, **59**, 86, 101
trees
  and carbon capture 65
  and emerging diseases 146–7
  and furniture 186
  planting schemes 153–6
  root systems 65
  see also deforestation; forests; re-forestation
tropics 42–3, 50, 155
Turkey 52

Uganda 102
Ulian, Tiziana 178
United Kingdom 48, 164
United Nations (UN) 103, 210, 211
United Nations Global Compact 168
United States 165, 171, 191, 206
universe, expansion 25–6
urban environments 192–3
Utila vii–viii

vegetables 178–80, **179**
Velcro 168–9
Venezuela 7
Victoria herbarium, Australia 9
viruses 20, 145–6
Vohibola littoral rainforest 160–1

Walker, Kim 84
walking 198
Wallace, Alfred Russel 7, 8, 70–1, **71**
walrus 131
Washington Post (newspaper) 99–100
wasps 36–7, 92
waste **150**, 167–8
water supplies 24–5, 88, 155, **159**, 167, 190–1
wealth, inclusive measures of 170
weather, extreme events 135–6
Werneck, Fernanda 129
wetlands 114
whales 6, 162
wildfires 72–3, 115, 135–6
Wilkin, Paul 166
willow 93
Wilson, Edward O. 29, 40
wind turbines 189, 190
wolves 31–2, 92–4, **95**
woodpeckers 94
wool 183
working from home 198
workplaces 200–1
World Wide Fund for Nature 70

Xishuangbanna Tropical Botanical Garden 118

yam 166
Yellowstone National Park 92–4, **95**
Yucatán, Mexico 24

Zimbabwe 169
zinc 86
Zizka, Alexander 43
zoogeographical regions 70–1, **71**